Darko Dimitrov

Geometric Applications of Principal Component Analysis

Darko Dimitrov

Geometric Applications of Principal Component Analysis

Quality of PCA Bounding Boxes and Detecting Symmetry

Südwestdeutscher Verlag für Hochschulschriften

Impressum / Imprint

Bibliografische Information der Deutschen Nationalbibliothek: Die Deutsche Nationalbibliothek verzeichnet diese Publikation in der Deutschen Nationalbibliografie; detaillierte bibliografische Daten sind im Internet über http://dnb.d-nb.de abrufbar.

Alle in diesem Buch genannten Marken und Produktnamen unterliegen warenzeichen-, marken- oder patentrechtlichem Schutz bzw. sind Warenzeichen oder eingetragene Warenzeichen der jeweiligen Inhaber. Die Wiedergabe von Marken, Produktnamen, Gebrauchsnamen, Handelsnamen, Warenbezeichnungen u.s.w. in diesem Werk berechtigt auch ohne besondere Kennzeichnung nicht zu der Annahme, dass solche Namen im Sinne der Warenzeichen- und Markenschutzgesetzgebung als frei zu betrachten wären und daher von jedermann benutzt werden dürften.

Bibliographic information published by the Deutsche Nationalbibliothek: The Deutsche Nationalbibliothek lists this publication in the Deutsche Nationalbibliografie; detailed bibliographic data are available in the Internet at http://dnb.d-nb.de.

Any brand names and product names mentioned in this book are subject to trademark, brand or patent protection and are trademarks or registered trademarks of their respective holders. The use of brand names, product names, common names, trade names, product descriptions etc. even without a particular marking in this works is in no way to be construed to mean that such names may be regarded as unrestricted in respect of trademark and brand protection legislation and could thus be used by anyone.

Coverbild / Cover image: www.ingimage.com

Verlag / Publisher:
Südwestdeutscher Verlag für Hochschulschriften
ist ein Imprint der / is a trademark of
AV Akademikerverlag GmbH & Co. KG
Heinrich-Böcking-Str. 6-8, 66121 Saarbrücken, Deutschland / Germany
Email: info@svh-verlag.de

Herstellung: siehe letzte Seite /
Printed at: see last page
ISBN: 978-3-8381-3433-8

Copyright © 2012 AV Akademikerverlag GmbH & Co. KG
Alle Rechte vorbehalten. / All rights reserved. Saarbrücken 2012

Geometric Applications
of Principal Component Analysis

Darko Dimitrov

Preface

Bounding boxes are used in many applications for simplification of point sets or complex shapes. For example, in computer graphics, bounding boxes are used to maintain hierarchical data structures for fast rendering of a scene or for collision detection. Additional applications include those in shape analysis and shape simplification, or in statistics, for storing and performing range-search queries on a large database of samples.

A frequently used heuristic for computing a bounding box of a set of points is based on principal component analysis. The principal components of the point set define the axes of the bounding box. Once the axis directions are given, the dimension of the bounding box is easily found by the extreme values of the projection of the points on the corresponding axis. Computing a PCA bounding box of a discrete point set in $\mathbb{R}^d$ depends linearly on the number of points. The popularity of this heuristic, besides its speed, lies in its easy implementation and in the fact that usually PCA bounding boxes are tight-fitting.

In this book we investigate the quality of the PCA bounding boxes. We give bounds on the worst case ratio of the volume of the PCA bounding box and the volume of the minimum volume bounding box. We present examples of point sets in the plane, where the worst case ratio tends to infinity. In these examples some dense point clusters have a big influence on the directions of the principal components, and the resulting PCA bounding boxes have much larger volumes than the minimal ones. To avoid the influence of such non-uniform distributions of the point sets, we consider PCA bounding boxes for continuous sets, especially for the convex hulls of point sets, obtaining several variants of continuous PCA. For those variants, we give lower bounds in arbitrary dimension, and upper bounds in $\mathbb{R}^2$ and $\mathbb{R}^3$. To obtain the lower bounds, we exploit a relation between the perfect reflective symmetry and the principal components of point sets. Each of the upper bounds in $\mathbb{R}^2$ and $\mathbb{R}^3$ is obtained from two parameterized bounds. The first bound is general for all bounding boxes, while to obtain the second bound, we exploit some of the properties of PCA, combining them with ideas from discrete geometry and integral calculus.

The relation between the perfect reflective symmetry and the principal components of point sets, leads to a straightforward algorithm for computing the planes of symmetry of perfect and approximate reflective symmetric point sets. For the same purpose, we present an algorithm based on geometric hashing.

Acknowledgements. The work presented in this book was supervised by PD Dr. Klaus Kriegel. I would like to thank him for his extremely useful guidance and collaboration. The software, based on the theoretical results presented here, was developed together with Dr. Mathias Holst. It was a great pleasure to work together with such a skillful programmer.

Darko Dimitrov
Berlin, 10. August 2012

Contents

Preface		iii
1 Introduction		**1**
2 Preliminaries		**9**
2.1	Some Matrix Algebra Revision	9
2.2	Multivariate Analysis	16
2.3	Principal Component Analysis	21
3 Lower Bounds on PCA Bounding Boxes		**33**
3.1	Approximation factors	33
3.2	Continuous PCA	35
3.3	Lower Bounds	36
4 Upper Bounds on PCA Bounding Boxes		**45**
4.1	Upper Bounds in $\mathbb{R}^2$	47
4.2	An upper bound in $\mathbb{R}^3$	63
4.3	Open Problems	74
5 Closed-form Solutions for Continuous PCA		**77**
5.1	Evaluation of the Expressions for Continuous PCA	77
6 Experimental Results		**85**
6.1	Evaluation of Bounding Box Algorithms	86
6.2	Conclusion	95
6.3	Additional results	96

7 Reflective Symmetry - an Application of PCA   107
  7.1 Introduction and Related Work 107
  7.2 Geometric Hashing Approach 110
  7.3 PCA Approach 120

Bibliography   125

Chapter 1
Introduction

Principle component analysis (PCA), also known as *Karhunen-Loeve transform*, or *Hotelling transform*, is one of the oldest and best known techniques of multivariate data analysis. The central idea of PCA is to reduce the dimensionality of a data set represented by d interrelated variables, while retaining as much as possible of the variation presented in the data set. This reduction is achieved by representing the data set with respect to a new set of d variables (a new coordinate system). The new set of variables, the so-called *principle components* (PCs), are chosen such that they are uncorrelated, and they are ordered so that the first *few* retain most of the variation present in *all* of the original variables. For a graphical illustration of this reduction, we consider a simple example of a 2-dimensional point set P in Figure 1.1 (in statistical data analysis PCA is usually applied to higher dimensional point sets). Primarily, P is given with respect to two variables X_1 and X_2, where the origin is chosen at the center of gravity of P. On the right side in Figure 1.1, we have the same point set with respect to the coordinate system defined by the principal components PC_1 and PC_2. As one can observe, the variable PC_1 approximates the set of observations much better than any of the old variables X_1 and X_2 in the sense that the orthogonal projection of P onto the line along PC_1 has the maximal variance among all possible projections. As it will be shown in the next chapter, if the relationship between X_1 and X_2 (and therefore between PC_1 and PC_2) is linear,

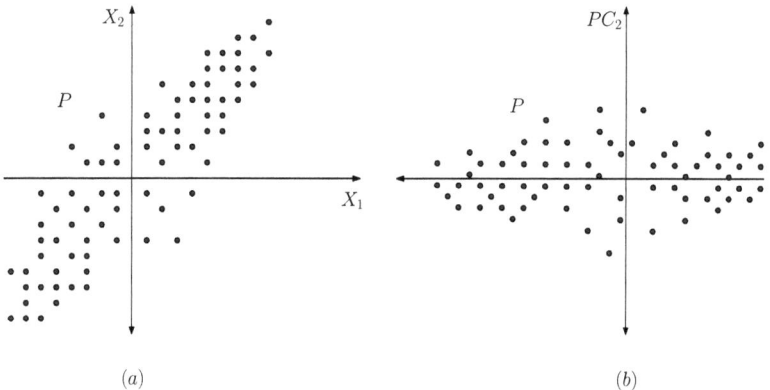

Figure 1.1: (a) Plot of 80 observations on two variables X_1 and X_2. (b) Plot of the same 80 observations from (a) with respect to their principal components PC_1 and PC_2.

than PC_1 will contain the whole variance of the observations, and we can discard PC_2 without losing any information.

The reduction of the dimensionality of the data set has many applications in various fields including computer vision, pattern recognition, visualization, data analysis, etc.

The computation of the principle components reduces to the solution of an eigenvalue-eigenvector problem for a positive-semidefinite symmetric matrix, which can be solved efficiently.

Geometric applications of PCA

Most of the applications of PCA are non-geometric in their nature. However, there are also a few purely geometric applications. A simple example is the estimation of the undirected normals of the points. That heuristic is based on the result by Pearson [38], who showed that the best-fitting line of the point set in a d-dimensional space is determined by the first principal component of the point set, and the direction of the last principal component is orthogonal to the best-fitting hyperplane of the point set. Then, for a given point cloud obtained from a smooth 2-manifold in $\mathbb{R}^3$ and a point p on the surface, we can estimate the undirected normal to the surface at p as follows: find all the points in a certain neighborhood of p and compute the principal components of those points. The last principal component is an estimate of the undirected normal at p. See Figure 1.2 for an illustration in $\mathbb{R}^2$.

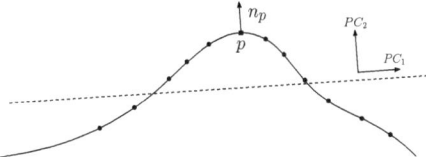

Figure 1.2: Estimation of the normal of the point p via principal components of its neighboring points.

In this book, we concentrate on the geometric properties of the PCA considering two geometric applications of it. First, we consider the problem of computing a bounding box of a point set in $\mathbb{R}^d$, and second, we consider the problem of detecting the perfect and approximate symmetry of a point set in $\mathbb{R}^d$.

Substituting sets of points or complex geometric shapes with their bounding boxes is motivated by many applications. For example, in computer graphics, it is used to maintain hierarchical data structures for fast rendering of a scene or for collision detection. Additional applications include those in shape analysis and shape simplification, or in statistics, for storing and performing range-search queries on a large database of samples.

Computing a minimum-area bounding rectangle of a set of n points in $\mathbb{R}^2$ can be done in $O(n \log n)$ time, for example with the rotating calipers algorithm [49]. O'Rourke [35] presented a deterministic algorithm, a rotating calipers variant in $\mathbb{R}^3$, for computing the minimum-volume bounding box of a set of n points in $\mathbb{R}^3$. His algorithm requires $O(n^3)$ time and $O(n)$ space. Barequet and Har-Peled [5] have contributed two $(1+\epsilon)$-approximation algorithms for the minimum-volume bounding box of point sets in $\mathbb{R}^3$, both with nearly linear complexity. The running times of their algorithms are $O(n + 1/\epsilon^{4.5})$ and $O(n \log n + n/\epsilon^3)$, respectively.

Numerous heuristics have been proposed for computing a box which encloses a given set of points. The simplest heuristic is naturally to compute the axis-aligned bounding box of the point set. Two-dimensional variants of this heuristic include the well-known *R-tree*, the *packed R-tree* [42], the R^*-*tree* [6], the R^+-*tree* [43], etc.

A frequently used heuristic for computing a bounding box of a set of points is based on PCA. The principal components of the point set define the axes of the bounding box. Once the directions of the axes are given, the dimension of the bounding box is easily found by the extreme values of the projection of the points on the corresponding axis.

Two distinguished applications of this heuristic are the OBB-tree [14] and the BOXTREE [4], hierarchical bounding box structures, which support efficient collision detection and ray tracing. Computing a bounding box of a set of points in $\mathbb{R}^2$ and $\mathbb{R}^3$ by PCA is simple and requires linear time. To avoid the influence of the distribution of the point set on the directions of the PCs, a possible approach is to consider the convex hull, or the boundary of the convex hull $CH(P)$ of the point set P. Thus, the complexity of the algorithm increases to $O(n \log n)$. The popularity of this heuristic, besides its speed, lies in

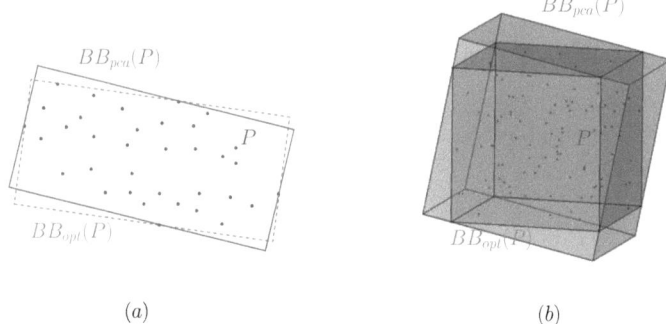

Figure 1.3: The minimum-area(volume) bounding box and the PCA bounding box of a point set P, (a) in $\mathbb{R}^2$, and (b) in $\mathbb{R}^3$.

its easy implementation and in the fact that usually PCA bounding boxes are tight-fitting, see Chapter 6 and [25] for some experimental results. Nevertheless, nothing was known about the approximation quality of the PCA bounding box algorithm in the worst case. Putting this problem into a striking phrase one could ask:

Is the computation of a bounding box by PCA only a heuristic (without any guarantees), or is a bounding box computed by PCA a constant factor approximation of the minimum-volume bounding box?

In this book we give answers to that question for discrete point sets in arbitrary dimension and for continuous convex point sets in $\mathbb{R}^2$ and $\mathbb{R}^3$.

Symmetry detection is an important problem with many applications in pattern recognition, computer vision and computational geometry. In the book, we use a relation between the perfect reflective symmetry and the principal components of shapes, a relation that is the basis for obtaining lower bounds for the quality of the PCA bounding boxes in this book. Namely, we show that any d-dimensional point set, symmetric with respect to a hyperplane H, has a principal component orthogonal to H. See Figure 1.4 for a 2-dimensional illustration. This relation leads to a simple and efficient algorithm for detecting hyperplanes of symmetry in arbitrary dimension. For the same purpose, we also present an algorithm based on geometric hashing.

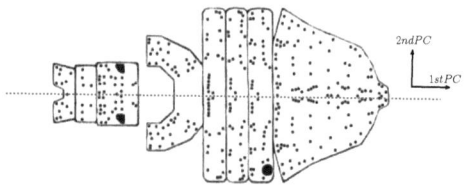

Figure 1.4: A pore pattern of a copepode (a microorganism that belongs to a group of small crustaceans found in the sea and nearly every freshwater habitat) and its axis of reflective symmetry.

Overview of the book

The main part of this book is dedicated to investigation of the quality of the PCA bounding boxes. We present bounds on the worst case ratio of the volume of the PCA bounding box and the volume of the minimum volume bounding box. The bounds presented here are the first results about the quality of the PCA bounding boxes. In addition, we consider the problem of detecting perfect and approximate reflective symmetry of a point set, applying PCA and geometric hashing approaches.

The structure and the contributions of the work presented here are as follows.

- In Chapter 2 we discuss preliminaries that will be used in several of the following chapters. We consider principal component analysis and present some important results about it.

- In Chapter 3 we present lower bounds on the approximation factor of PCA bounding boxes for point sets in arbitrary dimension. We present examples of discrete point sets in the plane, where the worst case ratio tends to infinity. As a consequence, it follows that the discrete setting can not lead to an approximation algorithm for the minimum-volume bounding box. To avoid the influence of the distribution of the point set on the directions of the principal components, we consider PCA bounding boxes for continuous sets, especially for the convex hull of a point set, obtaining several variants of continuous PCA. We investigate the quality of the bounding boxes obtained by the variants of continuous PCA related to the convex hull of a point set, giving lower bounds on the approximation factor in arbitrary dimension.

- In Chapter 4 we present upper bounds on the approximation factor of PCA bounding boxes for continuous convex point sets in $\mathbb{R}^2$ and $\mathbb{R}^3$.

- In Chapter 5 we consider the continuous version of PCA and give the closed form solutions for the case when the point set is a polyhedron or a polyhedral surface.

- In Chapter 6 we study the impact of the theoretical results on applications of several PCA variants in practice. We analyze the advantages and disadvantages of the different variants on realistic inputs, randomly generated inputs, and specially constructed (worst case) instances. Also, we evaluate and compare the performances of several existing bounding box algorithms we have implemented.

- In Chapter 7 we exploit a relation between the principal components and a hyperplane of symmetry of a perfect reflective-symmetric point set in arbitrary dimension. This relation implies a straightforward algorithm for detecting hyperplanes of symmetry. In addition, for the same purpose, we present an algorithm based on geometric hashing. 2D versions of both algorithms have been implemented and tested on real and synthetic data. The generation of the synthetic data is based on a probabilistic model, which additionally is used for a probabilistic analysis of the reliability of the geometric hashing algorithm.

Most of the results presented in this book have been published in [10], [11], [12], and [13].

Chapter 2
Preliminaries

In this chapter we collect known concepts and results which will be used several times within this book. The central part of this chapter is Section 2.3, where we consider principal component analysis and present some important results about it. To keep the presentation self-contained, in Section 2.1 and Section 2.2, we give an overview of some definitions and results from linear algebra and multivariate analysis. For those results which are normally not treated in undergraduate mathematics courses, we give additional comments and proofs. The results presented in this chapter are adapted from [8, 16, 34, 45].

2.1 Some Matrix Algebra Revision

Multivariate data consist of observations on several different variables for a number of individuals or objects. We denote the number of variables by d, and the number of individuals or objects by n. Thus in total we have $n \times d$ measurements. Let a_{rj} be the r-th observation of the j-th variable. The matrix whose element in the r-th row and j-th column is a_{rj}, is called the *data matrix* and is denoted by A. Thus

$$A = \begin{pmatrix} a_{11} & a_{12} & \cdots & a_{1d} \\ a_{21} & a_{22} & \cdots & a_{2d} \\ \vdots & \vdots & \ddots & \vdots \\ a_{n1} & a_{n2} & \cdots & a_{nd} \end{pmatrix}.$$

If A has n rows and d columns we say it is of order $n \times d$. The *transpose* of a matrix A is formed by interchanging the rows and columns, and we denote it by A^T. A matrix with column-order one is called a *column vector*. Thus

$$\mathbf{a} = \begin{pmatrix} a_1 \\ a_2 \\ \vdots \\ a_n \end{pmatrix}$$

is a column vector with n components. The data matrix can be seen as n row vectors, which we denote by $\mathbf{a}_1^T$ to $\mathbf{a}_n^T$, or as d column vectors, which we denote by $\mathbf{b}_1$ to $\mathbf{b}_d$. Thus

$$A = \begin{pmatrix} \mathbf{a}_1^T \\ \mathbf{a}_2^T \\ \vdots \\ \mathbf{a}_n^T \end{pmatrix} = (\mathbf{b}_1, \mathbf{b}_2, \ldots, \mathbf{b}_d)$$

where $\mathbf{a}_i^T$ denotes the *transpose* of $\mathbf{a}_i$. Note that the vectors are printed in bold type, and the matrices in ordinary type.

Also note that row vectors are points in a d-dimensional space, while the column vector are points in an n-dimensional space. When comparing variables, we compare column vectors.

2.1. SOME MATRIX ALGEBRA REVISION

A matrix is squared if the number of its rows equals to the number of its columns. A squared matrix is said to be *diagonal* if all its off-diagonal elements are zero. The *identity matrix*, denoted by I, is a diagonal matrix whose diagonal elements are all unity. The *trace* of a squared matrix A of order $d \times d$ is the sum of the diagonal terms, namely $\sum_{i=1}^{d} a_{ii}$, and will be denoted by $\operatorname{tr}(A)$.

A *determinant* of a squared matrix A is defined as

$$\det(A) = \sum \operatorname{sgn}(\tau)\, a_{1\tau(1)} \ldots a_{d\tau(d)},$$

where the summation is taken over all permutations τ of $(1, 2, \ldots, d)$, and $\operatorname{sgn}(\tau)$ denotes the *signature* of the permutation τ: $+1$ or -1, depending on whether τ can be written as the product of an even or odd number of transpositions.

A squared matrix A is *non-singular* if $\det(A) \neq 0$; otherwise it is singular.

The *inverse matrix* of matrix A is the unique matrix A^{-1} satisfying

$$AA^{-1} = A^{-1}A = I.$$

The inverse exists if and only if A is non-singular, that is, if and only if $\det(A) \neq 0$.

A set of vectors $\mathbf{x_1}, \ldots, \mathbf{x_d}$ is said to be *linearly dependent* if there exist constants $c_1, \ldots, c_d$ which are not all zero, such that

$$\sum_{i=1}^{d} c_i \mathbf{x_i} = \mathbf{0}.$$

Otherwise the vectors are said to be *linearly independent*. This definition leads on the idea of a *rank* of a matrix, which is defined as the maximum number of rows which are linearly independent (or equivalently as the maximum number of columns which are linearly independent). In other words, the rank is the dimension of the subspace spanned by vectors consisting of all the rows (or all the columns). We denote by $rank(A)$ the rank of matrix A. The following relations hold:

$$\begin{aligned} rank(A) &= rank(A^T) \\ &= rank(AA^T) \\ &= rank(A^TA) \end{aligned} \quad (2.1)$$

and
$$rank(A) = rank(BA) = rank(AC), \quad (2.2)$$
for all non-singular matrices B, C of appropriate order.

Orthogonality. Two vectors $\mathbf{x}, \mathbf{y}$ of order $d \times 1$ are said to be *orthogonal* if
$$\mathbf{x}^T\mathbf{y} = 0.$$
They are said to be *orthonormal* if they are orthogonal and
$$\mathbf{x}^T\mathbf{x} = \mathbf{y}^T\mathbf{y} = 1.$$
A squared matrix B is said to be *orthogonal* if
$$B^TB = BB^T = I$$
so that the rows (columns) of B are orthonormal. It is clear that B must be non-singular with
$$B^{-1} = B^T.$$

A transformation from a $d \times 1$ vector $\mathbf{x}$ to an $n \times 1$ vector $\mathbf{y}$ given by
$$\mathbf{y} = A\mathbf{x} + \mathbf{b}, \quad (2.3)$$
where A is an $n \times d$ matrix and $\mathbf{b}$ is an $n \times 1$ vector, is called a *linear transformation*. For $n = d$ the transformation is called *non-singular* if A is non-singular, and in that case the inverse transformation is
$$\mathbf{x} = A^{-1}(\mathbf{y} - \mathbf{b}).$$

2.1. SOME MATRIX ALGEBRA REVISION

An *orthogonal transformation* is defined by

$$\mathbf{y} = A\mathbf{x}, \qquad (2.4)$$

where A is an orthogonal matrix. Geometrically, an orthogonal matrix represents a linear transformation which consists of a rigid rotation, plus maybe reflection, since it preserves distances and angles. The determinant of an orthogonal matrix is ± 1. If the determinant is $+1$, the corresponding transformation is a pure rotation, while if the determinant is -1, the corresponding transformation involves in addition a reflection.

Quadratic forms and definiteness. A *quadratic form* in d variables, $x_1, \ldots, x_d$ is a function consisting of all possible second-order terms, namely

$$a_{11}x_1^2 + \cdots + a_{dd}x_d^2 + a_{12}x_1x_2 + \cdots + a_{d-1,d}x_{d-1}x_d = \sum_{1 \leq i,j \leq d} a_{ij}x_ix_j.$$

This can be conveniently written as $\mathbf{x}^T A \mathbf{x}$, where $\mathbf{x}^T = [x_1, \ldots, x_d]$. The matrix A is usually taken to be symmetric. A squared matrix A and its associated quadratic form is called:

- *positive definite* if $\mathbf{x}^T A \mathbf{x} > 0$ for every $\mathbf{x} \neq \mathbf{0}$;
- *positive semidefinite* if $\mathbf{x}^T A \mathbf{x} \geq 0$ for every $\mathbf{x}$.

Positive definite quadratics forms have matrices of full rank and can be represented as

$$A = QQ^T \qquad (2.5)$$

where Q is non-singular. Then $\mathbf{y} = Q^T \mathbf{x}$ transforms the quadratic form $\mathbf{x}^T A \mathbf{x}$ to the reduced form $y_1^2 + \cdots + y_d^2$ which only involves squared terms.

If A is a positive semidefinite of rank $m(< d)$, then A can also be expressed in the form of Equation (2.5), but with a matrix Q of order $d \times m$ which is of rank m. This is sometimes called the *Young-Householder factorization* of A.

Eigenvalues and eigenvectors. If Σ is a quadratic matrix of order $d \times d$, then
$$q(\lambda) = \det(\Sigma - \lambda I) \qquad (2.6)$$
is a d-th order polynomial in λ. It is called the *characteristic polynomial* of Σ. The d roots of $q(\lambda)$, $\lambda_1, \lambda_2, \ldots, \lambda_d$, possibly complex numbers, are called *eigenvalues* of Σ. Some of the λ_i will be equal if $q(\lambda)$ has multiple roots. To each eigenvalue λ_i, there corresponds a vector $\mathbf{c}_i$, called an eigenvector, such that
$$\Sigma \mathbf{c}_i = \lambda_i \mathbf{c}_i. \qquad (2.7)$$
The eigenvectors are not unique as they contain an arbitrary scale factor, and thus, they are usually normalized so that $\mathbf{c}_i^T \mathbf{c}_i = 1$. When there are equal eigenvalues, the corresponding eigenvectors can, and will, be chosen to be orthonormal.

If $\mathbf{x}$ and $\mathbf{y}$ are eigenvectors for λ_i and $\alpha \in \mathbb{R}$, then $\mathbf{x} + \mathbf{y}$ and $\alpha \mathbf{x}$ are also eigenvectors for λ_i. Thus, the set of all eigenvectors for λ_i forms a subspace which is called the *eigenspace* of Σ for λ_i. The maximal number of independent eigenvectors of the eigenspace determines the *dimension* of the eigenspace.

Some useful properties are as follows:.

(a)
$$\sum_{i=1}^{d} \lambda_i = \operatorname{tr}(\Sigma); \qquad (2.8)$$

(b)
$$\prod_{i=1}^{d} \lambda_i = \det(\Sigma); \qquad (2.9)$$

2.1. SOME MATRIX ALGEBRA REVISION

(c) If Σ is real symmetric matrix, then its eigenvalues and eigenvectors are real;

(d) If, further, Σ is positive definite, then all the eigenvalues are strictly positive;

(e) If Σ is positive semidefinite of rank m ($< d$), then Σ has m positive and $(d - m)$ zero eigenvalues;

(f) For two different eigenvalues, the corresponding normalized eigenvector are orthonormal;

(g) If we form a $d \times d$ matrix C, whose i-th column is the normalized eigenvector $\mathbf{c}_i$, then $C^T C = I$ and

$$C^T \Sigma C = \Lambda \tag{2.10}$$

where Λ is a diagonal matrix whose diagonal elements are $\lambda_1, \ldots, \lambda_d$. This is called the *canonical reduction* of Σ.

The matrix C is a quadratic form of Σ to the reduced form which only involves squared terms. Writing $\mathbf{x} = C\mathbf{y}$ we have

$$\begin{aligned} \mathbf{x}^T \Sigma \mathbf{x} &= \mathbf{y}^T C^T \Sigma C \mathbf{y} \\ &= \mathbf{y}^T \Lambda \mathbf{y} \\ &= \lambda_1 y_1^2 + \cdots + \lambda_m y_m^2 \end{aligned} \tag{2.11}$$

where $m =$ rank (Σ).

From Equation (2.10) we may also write

$$\Sigma = C \Lambda C^T = \lambda_1 \mathbf{c}_1 \mathbf{c}_1^T + \cdots + \lambda_m \mathbf{c}_m \mathbf{c}_m^T. \tag{2.12}$$

This is called the *spectral decomposition* of Σ.

Differentiation with respect to vectors. Suppose we have a differentiable function of d variables, say $f(x_1, \ldots, x_d)$. The notation $\partial f/\partial \mathbf{x}$

will be used to denote a column vector whose i-th component is the partial derivation $\partial f/\partial x_i$.

Suppose that the function is the quadratic form $\mathbf{x}^T\Sigma\mathbf{x}$, where Σ is a $d \times d$ symmetric matrix. Then it is straightforward to show that

$$\frac{\partial f}{\partial \mathbf{x}} = 2\Sigma\mathbf{x}. \tag{2.13}$$

2.2 Multivariate Analysis

2.2.1 Means, variances, covariances, and correlations

Let $\mathbf{X}$ be a random vector consisting of d random variables. Random vectors will be printed with capital letters in bold type. In this section we will present quantities that summarize a probability distribution of $\mathbf{X}$. In the univariate case, it is often done by giving the first two moments, namely the mean and the variance (or its square root, the standard deviation). To summarize multivariate distributions, we need to find the mean and variance of the d variables, together with a measure of the way each pair of variables is related. The latter target is achieved by calculating a set of quantities called *covariances*, or their standardized counterparts called *correlations*.

Means. The mean vector $\boldsymbol{\mu}^T = [\mu_1, \ldots \mu_d]$ such that

$$\mu_i = E(X_i) = \sum x P_i(x) \tag{2.14}$$

is the mean of the i-th component of $\mathbf{X}$. Here $P_i(x)$ denotes the (marginal) probability distribution of X_i. This definition is given for the case where X_i is discrete. If X_i is continuous, then

$$E(X_i) = \int_{-\infty}^{\infty} x f_i(x) dx, \tag{2.15}$$

where $f_i(x)$ is the probability density function of X_i.

2.2. MULTIVARIATE ANALYSIS

Variances. The variance of the i-th component of $\mathbf{X}$ is given by

$$\begin{aligned} \mathrm{var}(X_i) &= E[(X_i - \mu_i)^2] \\ &= E(X_i^2) - \mu_i^2. \end{aligned} \tag{2.16}$$

This is usually denoted by σ_i^2 in the univariate case, but in order to tie in with the covariance notation given below, we denote it by σ_{ii} in the multivariate case.

Covariances. The covariance of two variables X_i and X_j is defined by

$$\mathrm{cov}(X_i, X_j) = E[(X_i - \mu_i)(X_j - \mu_j)]. \tag{2.17}$$

Thus, it is the product moment of the two variables about their respective means. In particular, if $i = j$, we note that the covariance of a variable with itself is simply the variance of the variable. The covariance of X_i ad X_j is usually denoted by σ_{ij}. Thus, for $i = j$, σ_{ii} denotes the variance of X_i. Equation 2.17 is often written in the equivalent alternative form

$$\sigma_{ij} = E[X_i X_j] - \mu_i \mu_j. \tag{2.18}$$

The covariance matrix. Given d variables, there are d variances and $d(d-1)/2$ covariances, and all these quantities are second moments. It is often useful to present these quantities in a symmetric $d \times d$ matrix, denoted by Σ, whose (i,j)-th element is σ_{ij}. Thus,

$$\Sigma = \begin{pmatrix} \sigma_{11} & \sigma_{12} & \cdots & \sigma_{1d} \\ \sigma_{21} & \sigma_{22} & \cdots & \sigma_{2d} \\ \vdots & \vdots & \ddots & \vdots \\ \sigma_{d1} & \sigma_{d2} & \cdots & \sigma_{dd} \end{pmatrix}. \tag{2.19}$$

The matrix is variously called the *dispersion* matrix, the *variance-covariance* matrix, or simply the *covariance* matrix, and we will use the latter term. The diagonal terms of Σ are the variances, while the off-diagonal terms are the covariances.

Using Equations (2.17) and (2.18), we can express Σ in two alternative useful forms, namely

$$\begin{aligned} \Sigma &= E[(\mathbf{X} - \boldsymbol{\mu})(\mathbf{X} - \boldsymbol{\mu})^T] \\ &= E[\mathbf{X}\mathbf{X}^T] - \boldsymbol{\mu}\boldsymbol{\mu}^T. \end{aligned} \qquad (2.20)$$

Linear combination. Perhaps the main use of covariances is as a stepping stone to the calculations of correlations (see below), but they are also useful for a variety of other purposes. Later in Section 2.3, we will consider the variance of a linear combination of the components of $\mathbf{X}$. Consider the general linear combination

$$Y = \mathbf{a}^T \mathbf{X}$$

where $\mathbf{a}^T = [a_1, \ldots, a_d]$ is a vector of constants. Then Y is a univariate random variable. Its mean is clearly given by

$$E(Y) = \mathbf{a}^T \boldsymbol{\mu} \qquad (2.21)$$

while its variance is given by

$$\mathrm{var}(Y) = E[\{\mathbf{a}^T(\mathbf{X} - \boldsymbol{\mu})\}^2]. \qquad (2.22)$$

As $\mathbf{a}^T(\mathbf{X} - \boldsymbol{\mu})$ is a scalar and therefore equal to its transpose, we can express $\mathrm{var}(Y)$ in terms of Σ, using Equation (2.20), as

$$\begin{aligned} \mathrm{var}(Y) &= E[\mathbf{a}^T(\mathbf{X} - \boldsymbol{\mu})(\mathbf{X} - \boldsymbol{\mu})^\mathbf{T}\mathbf{a}] \\ &= \mathbf{a}^T E[(\mathbf{X} - \boldsymbol{\mu})(\mathbf{X} - \boldsymbol{\mu})^\mathbf{T}]\mathbf{a} \\ &= \mathbf{a}^T \Sigma \mathbf{a}. \end{aligned} \qquad (2.23)$$

Correlations. Although covariances are useful for many mathematical purposes, they are rarely used as descriptive statistics. If two variables are related in a linear way, then the covariance will be positive or negative depending on whether the relationship has a positive or negative slope. But the size of the coefficient is difficult to interpret because it depends on the units in which the two variables are measured. Thus the covariance is often standardized by dividing by the product of the standard deviations of the two variables to give a quantity called the *correlation coefficient*. The correlation between variables X_i and X_j will be denoted by ρ_{ij}, and is given by

$$\rho_{ij} = \sigma_{ij}/\sigma_i\sigma_j \tag{2.24}$$

where σ_i and σ_j denote the standard deviations of X_i and X_j. It can be shown that ρ_{ij} is always a value between -1 and $+1$.

The correlation coefficient provides a measure of the linear association between two variables. The coefficient is positive if the relationship between the two variables has a positive slope so that the 'high' values of one variable tend to go with 'high' values of the other variable. Conversely, the coefficient is negative if the relationship has a negative slope.

If two variables are independent then their covariance, and their correlation, are zero. But it is important to note that the converse of this statement is not true. Here is an example: Suppose that the random variable X is uniformly distributed on the interval from -1 to 1, and $Y = X^2$. Then, Y is completely determined by X, so that X and Y are dependent, but their correlation is zero. This emphasizes the fact that the correlation coefficient may be misleading if the relationship between two variables is non-linear. However, if the two variables follow a bivariate normal distribution, then it turns out that zero correlation *does* imply independence. A detailed explanation and further results about the relation between the correlation and independence of the random variables can be found, for example, in

[45].

The correlation matrix. From the definition of the correlation, it follows that for given d variables, there are $d(d-1)/2$ distinct correlations. A $d \times d$ matrix, whose (i,j)-th element is defined to be ρ_{ij} is called the correlation matrix and will be denoted by $\mathcal{P}$ (capital Greek letter rho). A correlation matrix is symmetric with the diagonal terms all equal unity.

In order to relate the covariance matrix and correlation matrix, let us define a $d \times d$ diagonal matrix D, whose diagonal terms are the standard deviations of the components of $\mathbf{X}$, so that

$$D = \begin{pmatrix} \sigma_1 & 0 & \ldots & 0 \\ 0 & \sigma_2 & \ldots & 0 \\ \vdots & \vdots & \ddots & \vdots \\ 0 & 0 & \ldots & \sigma_d \end{pmatrix}. \quad (2.25)$$

Then the covariance matrix and correlation matrices are related by

$$\Sigma = D\mathcal{P}D$$

or

$$\mathcal{P} = D^{-1}\Sigma D^{-1} \quad (2.26)$$

where the diagonal terms of the matrix D^{-1} are the reciprocals of the respective standard deviations.

The rank of Σ and $\mathcal{P}$. We complete this section with a discussion of the matrix properties of Σ and $\mathcal{P}$, and in particular of their rank.

Firstly, we show that both Σ and $\mathcal{P}$ are positive semidefinite. As any variance must be non-negative, we have that

$$\text{var}(\mathbf{a}^T\mathbf{X}) \geq 0 \quad \text{for every } \mathbf{a}.$$

But $\text{var}(\mathbf{a}^T\mathbf{X}) = \mathbf{a}^T\Sigma\mathbf{a}$, and so Σ must be semidefinite. We also note that Σ is related to $\mathcal{P}$ by Equation (2.26), where D is non-singular, and so it follows that $\mathcal{P}$ is also positive semidefinite.

Because D is non-singular, we may also use Equations (2.26) and (2.2) to show that the rank of $\mathcal{P}$ is the same as the rank of Σ. This rank must be less than or equal d.

If Σ (and hence $\mathcal{P}$) has rank d, then Σ ($\mathcal{P}$) is positive definite, as in this case, $\text{var}(\mathbf{a}^T\mathbf{X})$ is strictly greater than zero for every $\mathbf{a} \neq \mathbf{0}$. But if $\text{rank}(\Sigma) < d$, then Σ ($\mathcal{P}$) is singular, and this indicates a linear constraint on the components of $\mathbf{X}$. This means that there exists a vector $\mathbf{a} \neq \mathbf{0}$ such that $\text{var}(\mathbf{a}^T\mathbf{X}) = \mathbf{a}^T\Sigma\mathbf{a}$ is zero, indicating that Σ is positive semidefinite rather than positive definite.

When $\text{rank}(\Sigma) < d$, the components of $\mathbf{X}$ are sometimes said to be 'linearly dependent', using this term in its algebraic sense. However, statisticians often use this term to mean a linear relationship between the *expected values* of the random variables. It needs to be emphasized that a constraint of the latter type will generally *not* produce a singular Σ. If two variables are correlated, it does not mean that one of them is redundant, although if the correlation is very high then one of them may be 'nearly redundant' and the covariance matrix will be 'nearly singular'.

2.3 Principal Component Analysis

2.3.1 Motivation

The central idea and motivation of principal component analysis (abbreviated to PCA) is to reduce the dimensionality of a point set by identifying *the most significant directions (principal components)*. Let $P = \{\mathbf{p_1}, \mathbf{p_2}, \ldots, \mathbf{p_n}\}$ be a set of vectors (points) in $\mathbb{R}^d$, and

$\boldsymbol{\mu} = (\mu_1, \mu_2, \ldots, \mu_d) \in \mathbb{R}^d$ be the center of gravity of P. For $1 \leq k \leq d$, we use p_{ik} to denote the k-th coordinate of the vector $\mathbf{p_i}$. Given two vectors $\mathbf{u}$ and $\mathbf{v}$, we use $\langle \mathbf{u}, \mathbf{v} \rangle$ to denote their inner product. For any unit vector $\mathbf{v} \in \mathbb{R}^\mathbf{d}$, the *variance of P in direction v* is

$$\text{var}(P, \mathbf{v}) = \frac{1}{n} \sum_{i=1}^{n} \langle \mathbf{p_i} - \boldsymbol{\mu}, \mathbf{v} \rangle^2. \tag{2.27}$$

The most significant direction corresponds to the unit vector $\mathbf{v_1}$ such that $\text{var}(P, \mathbf{v_1})$ is maximum. In general, after identifying the j most significant directions $\mathbf{v_1}, \ldots, \mathbf{v_j}$, the $(j+1)$-st most significant direction corresponds to the unit vector $\mathbf{v_{j+1}}$ such that $\text{var}(P, \mathbf{v_{j+1}})$ is maximum among all unit vectors perpendicular to $\mathbf{v_1}, \mathbf{v_2}, \ldots, \mathbf{v_j}$.

From the multivariate analysis point of view, we can consider P as a sample of points that represents a d-dimensional vector of random variables $\mathbf{X}^T = [X_1, X_2, \ldots X_d]$. Namely, to each coordinate of the points corresponds one random variable. As it will be shown in the next subsection principal components are uncorrelated *linear* combinations of the original variables of $\mathbf{X}$, and are derived in decreasing order of importance so that, for example, the first principal component accounts for as much possible of the variation in the original data. The transformation is in fact an *orthogonal rotation* in d-space. The technique of finding this transformation is called principal component analysis. PCA originated in the work by Karl Pearson [38] around the turn of the 20th century, and was further developed in the 1930s by Harold Hotelling [20] using the approach described in the next subsection.

The usual objective of the analysis is to study if the first few components account for most of the variation in the original data. If they do, then it is argued that the effective dimensionality of the problem is less than d. In order words, if some of the original variables are highly correlated, they are effectively 'saying the same thing' and there may be near-linear constraints on the variables. In this case it is hoped that the first few components will be intuitively meaningful, will help us understand the data better, and will be useful in subsequent analysis where we can operate with a smaller number of variables. The reduction of the complexity (dimensionality), we have illustrated in the introduction (Figure 1.1) on an unrealistic, but simple, case where $d = 2$.

We note that *PCA* is a statistical technique which does not require the user to specify an underlying statistical model to explain the 'error' structure. In particular, no assumption is made about the probability distribution of the original variables, though more meaning can generally be given to the components in the case where the observations are assumed to be multivariate normal [16].

2.3.2 Derivation of principal components

Suppose that $\mathbf{X}$ is a d-dimensional random vector with mean $\boldsymbol{\mu}$ and covariance matrix Σ. Our problem is to find a new set of variables, say $Y_1, Y_2, \ldots, Y_d$ which are uncorrelated and whose variances decrease from first to last. Each Y_j is taken to be a linear combination of the X's, so that

$$\begin{aligned} Y_j &= a_{1j}X_1 + a_{2j}X_2 + \cdots + a_{pj}X_d \\ &= \mathbf{a}_j^T \mathbf{X} \end{aligned} \quad (2.28)$$

where $\mathbf{a}_j^T = [a_{1j}, a_{2j}, \ldots, a_{pj}]$ is a vector of constants. Equation (2.28) contains an arbitrary scale factor. We therefore impose the condition that $\mathbf{a}_j^T \mathbf{a}_j = \sum_{k=1}^{d} a_{kj}^2 = 1$. We will call such a linear transformation a *standardized linear transformation*. We shall see that this particular normalization procedure ensures that the overall transformation is orthogonal - in other words, that distances in d-space are preserved.

The first principal component, Y_1, is found by choosing $\mathbf{a}_1$ so that Y_1 has the largest possible variance. In other words, we choose $\mathbf{a}_1$ so as to maximize the variance of $\mathbf{a}_1^T \mathbf{X}$ subject to the constraint that $\mathbf{a}_1^T \mathbf{a}_1 = 1$. This approach, originally suggested by Harold Hotelling [20], gives equivalent results to that of Karl Pearson [38], which finds the hyperplane in d-space such that the total sum of squared perpendicular distances from the point to the hyperplane is minimized.

The second principal component is found by choosing $\mathbf{a}_2$ so that Y_2 has the largest possible variance for all combinations of the form of Equation (2.28) which are uncorrelated with Y_1. Similarly, we derive $Y_3, \ldots, Y_d$, so as to be uncorrelated and to have decreasing variance.

We begin by finding the first component. We want to choose $\mathbf{a}_1$ so as to maximize the variance of Y_1 subject to normalization constraint that $\mathbf{a}_1^T \mathbf{a}_1 = 1$. Now

$$\begin{aligned} \operatorname{var}(Y_1) &= \operatorname{var}(\mathbf{a}_1^T \mathbf{X}) \\ &= \mathbf{a}_1^T \Sigma \mathbf{a}_1 \end{aligned} \qquad (2.29)$$

using Equation (2.23). Thus we take $\mathbf{a}_1^T \Sigma \mathbf{a}_1$ as our objective function.

The standard procedure for maximizing a function of several variables subject to one or more constraints is the method of Lagrange multipliers. With just one constraint, this method uses the fact that the stationary points of a differentiable function of d variables, say $f(x_1, \ldots, x_d)$, subject to a constraint $g(x_1, \ldots x_d) = c$, are such that there exists a number λ, called the *Lagrange multiplier*, such that

$$\frac{\partial f}{\partial x_i} - \lambda \frac{\partial g}{\partial x_i} = 0 \qquad i = 1, \ldots, d \qquad (2.30)$$

at the stationary points. These d equations, together with the constraints, are sufficient to determine the coordinates of the stationary points and the corresponding value of λ. Further investigations are

2.3. PRINCIPAL COMPONENT ANALYSIS

needed to see if a stationary point is a maximum, minimum or saddle point. It is helpful to form a new function $L(\mathbf{x})$, such that

$$L(\mathbf{x}) = f(\mathbf{x}) - \lambda[g(\mathbf{x}) - c]$$

where the term in the square brackets is of course zero. Then the set of equations in (2.30) may be written simply as

$$\frac{\partial L}{\partial \mathbf{x}} = \mathbf{0}$$

using the definition given in Section 2.1. Applying this method to our problem, we write

$$L(\mathbf{a}_1) = \mathbf{a}_1^T \Sigma \mathbf{a}_1 - \lambda(\mathbf{a}_1^T \mathbf{a}_1 - 1).$$

Then, using Equation 2.13, we have

$$\frac{\partial L}{\partial \mathbf{a}_1} = 2\Sigma \mathbf{a}_1 - 2\lambda \mathbf{a}_1.$$

Setting this equal to $\mathbf{0}$, we have

$$(\Sigma - \lambda I)\mathbf{a}_1 = \mathbf{0}, \qquad (2.31)$$

where I is the identity matrix of order $d \times d$. We now come to the crucial step in the argument. If Equation (2.31) is to have a solution for $\mathbf{a}_1$, other than the null vector, then $(\Sigma - \lambda I)$ must be a singular matrix. Thus λ must be chosen so that

$$\det(\Sigma - \lambda I) = 0.$$

Thus a non-zero solution for Equation (2.31) exists if and only if λ is an eigenvalue of Σ. But Σ will generally have d eigenvalues, which must all be nonnegative as Σ is positive semidefinite. Let $\lambda_1 \geq \lambda_2 \geq \cdots \geq \lambda_d \geq 0$ be the eigenvalues of Σ.

In the case where some of the eigenvalues are equal, there is no unique way of choosing the corresponding eigenvectors. Then, the eigenvectors associated with multiple roots will be chosen to be orthogonal.

Which eigenvalue shall we choose to determine the first principal component? Now,

$$\begin{aligned} \text{var}(\mathbf{a}_1^T \mathbf{X}) &= \mathbf{a}_1^T \Sigma \mathbf{a}_1 \\ &= \mathbf{a}_1^T \lambda \mathbf{a}_1 \quad \text{using Equation (2.31)} \\ &= \lambda. \end{aligned}$$

As we want to maximize this variance, we choose λ to be the *largest* eigenvalue, namely λ_1. Then, using Equation (2.31), the principal component, $\mathbf{a}_1$, which we are looking for must be the eigenvector of Σ corresponding to the largest eigenvalue.

The second principal component, namely $Y_2 = \mathbf{a}_2^T \mathbf{X}$, is obtained by an extension of the above argument. In addition to the scaling constraint that $\mathbf{a}_2^T \mathbf{a}_2 = 1$, we now have a second constraint that Y_2 should be uncorrelated with Y_1. Now,

$$\begin{aligned} \text{cov}(Y_2, Y_1) &= \text{cov}(\mathbf{a}_2^T \mathbf{X}, \mathbf{a}_1^T \mathbf{X}) \\ &= E(\mathbf{a}_2^T (\mathbf{X} - \mu)(\mathbf{X} - \mu)^T \mathbf{a}_1) \\ &= \mathbf{a}_2^T \Sigma \mathbf{a}_1. \end{aligned} \quad (2.32)$$

We require this to be zero. But since $\Sigma \mathbf{a}_1 = \lambda_1 \mathbf{a}_1$, an equivalent simpler condition is that $\mathbf{a}_2^T \mathbf{a}_1 = 0$. In order words, $\mathbf{a}_1$ and $\mathbf{a}_2$ should be orthogonal.

In order to maximize the variance of Y_2, namely $\mathbf{a}_2^T \Sigma \mathbf{a}_2$, subject to the two constraints, we need to introduce two Lagrange multipliers, which we will denote by λ and δ, and consider the function

$$L(\mathbf{a}_2) = \mathbf{a}_2^T \Sigma \mathbf{a}_2 - \lambda(\mathbf{a}_2^T \mathbf{a}_2 - 1) - \delta \mathbf{a}_2^T \mathbf{a}_1.$$

At the stationary point(s) we must have

$$\frac{\partial L}{\partial \mathbf{a}_2} = 2(\Sigma - \lambda I)\mathbf{a}_2 - \delta \mathbf{a}_1 = \mathbf{0}. \quad (2.33)$$

2.3. PRINCIPAL COMPONENT ANALYSIS

If we premultiply this equation by $\mathbf{a}_1^T$, we obtain

$$\mathbf{a}_1^T \Sigma \mathbf{a}_2 - \delta = 0$$

since $\mathbf{a}_1^T \mathbf{a}_2 = 0$. But from Equation (2.32), we also require $\mathbf{a}_1^T \Sigma \mathbf{a}_2$ to be zero, so that δ is zero at the stationary point(s). Thus Equation (2.33) becomes

$$(\Sigma - \lambda I)\mathbf{a}_2 = \mathbf{0}.$$

With a little thought, we see that this time we choose λ to be the *second* largest eigenvalue of Σ, and $\mathbf{a}_2$ to be the corresponding eigenvector.

Continuing this argument, the j-th principal component turns out to be the eigenvector associated with the j-th largest eigenvalue.

2.3.3 Definition and some properties of the principal components

Now, we give the definition of the principal component transformation of a random vector $\mathbf{X}$ for a general dimension d.

Definition 2.1. *If $\mathbf{X}$ is a random vector with mean $\boldsymbol{\mu}$ and covariance Σ, then the principal component transformation is the transformation*

$$\mathbf{X} \to \mathbf{Y} = C^T(\mathbf{X} - \boldsymbol{\mu}), \tag{2.34}$$

where C is orthogonal, $C^T \Sigma C = \Lambda$ is a diagonal matrix, with diagonal elements $\lambda_1 \geq \lambda_2 \geq \cdots \geq \lambda_d \geq 0$. The strict positivity of the eigenvalues λ_i, is guaranteed if Σ is positive definite. This representation of Σ follows from its spectral decomposition (Equation (2.10)). The i-th principal component of $\mathbf{X}$ may be defined as the i-th element of the vector $\mathbf{Y}$, namely as

$$Y_i = \mathbf{c_i}(\mathbf{X} - \boldsymbol{\mu}). \tag{2.35}$$

Here c_i is the i-th column of C and may be called the i-th vector of principal component loadings. The function Y_d may be called the last principal component of $\mathbf{X}$.

In the sequel, we summarize some fundamental properties of principal components, which are immediately observed from the derivation and definition of principal components.

Proposition 2.1. *If $\mathbf{X}$ is a random vector with mean $\boldsymbol{\mu}$ and covariance Σ, and $\mathbf{Y}$ is as defined in (2.34), then*

(a) $E(Y_i) = 0$;

(b) $var(Y_i) = \lambda_i$;

(c) $cov(Y_i, Y_j) = 0$, $i \neq j$;

(d) $var(Y_1) \geq var(Y_2) \cdots \geq var(Y_d) \geq 0$;

(e) $\sum_{i=1}^{d} var(Y_i) = tr(\Sigma)$;

(f) $\prod_{i=1}^{d} var(Y_i) = det(\Sigma)$

Proof. (a)-(d) follow from Definition 2.1 and the properties of the expectation operator. (e) follows from (b) and the fact that $tr(\Sigma)$ is the sum of the eigenvalues (Equation (2.8)). (f) follows from (b) and the fact that $det(\Sigma)$ is the product of the eigenvalues (Equation (2.9)). $\square$

From the derivation of the principal components in the previous section follows the next important property.

Proposition 2.2. *No standardized linear combination of $\mathbf{X}$ has a variance larger than λ_1, the variance of the first principal component.*

Similarly, as in Proposition 2.2, it follows that the last principal component of $\mathbf{X}$ has a variance which is *smaller* than that of any other

2.3. PRINCIPAL COMPONENT ANALYSIS

standardized linear combination. The intermediate components have a maximal variance property given by the following proposition.

Proposition 2.3. *If $\alpha = \mathbf{a}^T \mathbf{X}$ is a standardized linear combination of $\mathbf{X}$ which is uncorrelated with the first k principal components of $\mathbf{X}$, then the variance of α is maximized when α is the $(k+1)$-st principal component of $\mathbf{X}$.*

Now, we give a geometric property of principal components. Let A be a positive definite matrix of order $d \times d$. Then

$$(\mathbf{x} - \boldsymbol{\alpha})^T A^{-1} (\mathbf{x} - \boldsymbol{\alpha}) = c^2 \tag{2.36}$$

represents an ellipsoid in d dimensions with center $\mathbf{x} = \boldsymbol{\alpha}$. On shifting the center to $\mathbf{x} = \mathbf{0}$, the equation becomes

$$\mathbf{x}^T A^{-1} \mathbf{x} = c^2. \tag{2.37}$$

Definition 2.2. *Let $\mathbf{x}$ be a point on the ellipsoid defined by (2.36) and let $f(\mathbf{x}) = \|\mathbf{x} - \boldsymbol{\alpha}\|$ denote the squared distance between $\mathbf{x}$ and $\boldsymbol{\alpha}$. A line through $\boldsymbol{\alpha}$ and $\mathbf{x}$ for which $\mathbf{x}$ is a stationary point of $f(\mathbf{x})$ is called a principal axis of the ellipsoid. The distance $\|\mathbf{x} - \boldsymbol{\alpha}\|$ is called the length of the principal semi-axis.*

Proposition 2.4. *Let $\lambda_1, \ldots, \lambda_d$ be the eigenvalues of A satisfying $\lambda_1 > \lambda_2 > \cdots > \lambda_d$. Suppose that $\mathbf{c_1}, \ldots, \mathbf{c_d}$ are the corresponding eigenvectors. For the ellipsoids (2.36) and (2.37), we have*

(a) The direction of the i-th principal axis coincides with $\mathbf{c_i}$.

(b) The length of the i-th principal semi-axis is $c\lambda_i^{1/2}$.

Proof. It is sufficient to prove the result for (2.37). The problem reduces to finding the stationary points of $f(\mathbf{x}) = \mathbf{x}^T \mathbf{x}$ subject to $\mathbf{x}$ lying on the ellipsoid $\mathbf{x}^T A^{-1} \mathbf{x} = c^2$. From (2.13), we have that

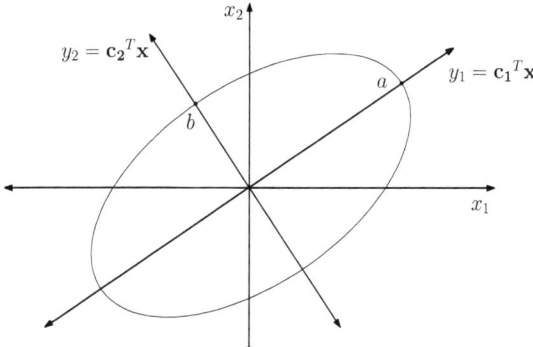

Figure 2.1: Ellipsoid $\mathbf{x}^T A^{-1}\mathbf{x} = 1$. Lines defined by y_1 and y_2 are the first and second principal axes, $\|\mathbf{a}\| = \lambda_1^{1/2}, \|\mathbf{b}\| = \lambda_2^{1/2}$.

the derivation of $\mathbf{x}^T A^{-1}\mathbf{x}$ is $2\mathbf{x}^T A^{-1}$. Thus a point $\mathbf{y}$ represents a direction tangent to the ellipsoid at $\mathbf{x}$ if $2\mathbf{y}^T A^{-1}\mathbf{x} = 0$.

The derivative of $f(\mathbf{x})$ is $2\mathbf{x}$ so the directional derivative of $f(\mathbf{x})$ in the direction $\mathbf{y}$ is $2\mathbf{y}^T\mathbf{x} = 0$; that is if

$$\mathbf{y}^T A^{-1}\mathbf{x} = 0 \Rightarrow \mathbf{y}^T\mathbf{x} = 0.$$

This condition is satisfied if and only if $A^{-1}\mathbf{x}$ is proportional to $\mathbf{x}$; that is if and only if $\mathbf{x}$ is an eigenvector of A^{-1}.

Setting $\mathbf{x} = \beta \mathbf{c_i}$ in (2.37) gives $\beta^2/\lambda = c^2$, so $\beta = c\lambda_i^{1/2}$. Thus, the theorem is proved. □

If we rotate the coordinate axes with the transformation $\mathbf{y} = C^T\mathbf{x}$, where C is obtained by the spectral decomposition of A ($A = C\Lambda C$), we find that (2.37) reduces to

$$\sum y_i^2/\lambda_i = c^2.$$

Figure 2.1 gives a pictorial representation.

With $A = I$, Equation (2.37) reduces to a hypersphere with $\lambda_1 = \cdots = \lambda_d = 1$ so that the λs are not distinct and the above theorem

2.3. PRINCIPAL COMPONENT ANALYSIS

fails; that is, the positions of c_i, $i = 1, \ldots d$ through the sphere are not unique and any rotation will suffice. In general, if $\lambda_i = \lambda_{i+1}$, the section of the ellipsoid is *circular* in the plane generated by c_i and c_{i+1}. Although we can construct two perpendicular axes for the common root, their position through the circle is not unique.

An immediate consequence of Proposition 2.4 follows.

Corollary 2.1. *Consider the family of d-dimensional ellipsoids*

$$\mathbf{X}^T \Sigma^{-1} \mathbf{X} = c^2. \qquad (2.38)$$

where $\mathbf{X}$ is a random vector with covariance matrix Σ. The principal components of $\mathbf{X}$ define the principal axes of these ellipsoids.

The properties of PCA, mentioned till now, were obtained from a known population covariance matrix Σ. In practice the covariance matrix Σ (or correlation matrix $\mathcal{P}$) is rarely known and hence the eigenvalues $\lambda_1, \ldots, \lambda_d$ and its corresponding eigenvectors must be estimated from the random sample (sample data matrix). However, in that case one can similarly obtain the results presented in this chapter. We conclude this chapter with the following property that is geometrically equivalent to the algebraic properties from Propositions 2.2 and 2.3.

Proposition 2.5. *Let $p_1, \ldots, p_n$ be observations in a d-dimensional space. A measure of 'goodness-of-fit' of a q-dimensional hyperplane to $p_1, \ldots, p_n$ can be defined as the sum of squared perpendicular distances of $p_1, \ldots, p_n$ from the hyperplane. This measure is minimized when the q-dimensional hyperplane is spanned by the first q principal components of $p_1, \ldots, p_n$.*

This property can be viewed as an alternative derivation of the principal components. Rather than adapting the algebraic definition of population principal components, given above in this section, there is an alternative geometric definition of sample principal components. They are defined as the linear functions (projections) of $\mathbf{p_1}, \ldots, \mathbf{p_n}$ that successively define hyperplanes of dimension $1, 2, \ldots, q, \ldots, (d-1)$ for which the sum of squared perpendicular distances of $\mathbf{p_1}, \ldots, \mathbf{p_n}$ from the hyperplane is minimized. This definition provides another way in which principal components can be interpreted as accounting for as much as possible of the total variation in the data, within lower-dimensional space. In fact, this is essentially the approach adopted by Pearson [38], although he concentrated on the two cases, where $q = 1$ and $q = d - 1$. Given a set of points in d-dimensional space, Pearson found the 'best-fitting line' and 'best-fitting hyperplane' in the sense of minimizing the sum of squared deviations of the points from the line or hyperplane. The best-fitting line determines the first principal component, although Pearson did not use this terminology, and the direction of the last principal component is orthogonal to the best-fitting hyperplane.

Chapter 3

Lower Bounds on the Quality of the PCA Bounding Boxes

In this and the next chapter we study the quality of the PCA bounding boxes, obtaining bounds on the worst case ratio of the volume of the PCA bounding box and the volume of the minimum-volume bounding box. We present examples of point sets in the plane, where the worst case ratio tends to infinity. To avoid the influence of the distribution of the point set on the directions of the PCs, we consider PCA bounding boxes for continuous sets, especially for the convex hull of a point set, obtaining several variants of continuous PCA. In this chapter, we investigate the quality of the bounding boxes obtained by the variants of continuous PCA related to the convex hull of a point set, giving lower bounds on the approximation factor in an arbitrary dimension.

3.1 Approximation factors

Given a point set $P \subseteq \mathbb{R}^d$ we denote by $BB_{pca}(P)$ the PCA bounding box of P and by $BB_{opt}(P)$ the bounding box of P with smallest possible volume. The ratio of the two volumes $\kappa_d(P) = \text{Vol}(BB_{pca}(P))/\text{Vol}(BB_{opt}(P))$ defines the approximation factor for P, and

$$\kappa_d = \sup \left\{ \kappa_d(P) \mid P \subseteq \mathbb{R}^d, \text{Vol}(CH(P)) > 0 \right\}$$

defines the general PCA approximation factor.

Since bounding boxes of a point set P (with respect to any orthog-

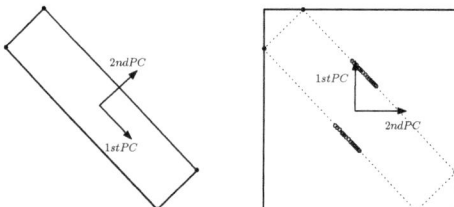

Figure 3.1: Four points and their PCA bounding-box (left). A dense collection of additional points significantly affect the orientation of the PCA bounding-box (right).

onal coordinate system) depend only on the convex hull of $CH(P)$, the construction of the covariance matrix should be based only on $CH(P)$ and not on the distribution of the points inside. Using the vertices, i.e., the 0-dimensional faces of $CH(P)$ to define the covariance matrix Σ we obtain a bounding box $BB_{pca(d,0)}(P)$. We denote by $\kappa_{d,0}(P)$ the approximation factor for the given point set P and by

$$\kappa_{d,0} = \sup \left\{ \kappa_{d,0}(P) \mid P \subseteq \mathbb{R}^d, \operatorname{Vol}(CH(P)) > 0 \right\}$$

the approximation factor in general. The example in Figure 3.1 shows that $\kappa_{2,0}(P)$ can be arbitrarily large if the convex hull is a thin, slightly "bulged rectangle", with a lot of additional vertices in the middle of the two long sides. Since this construction can be lifted into higher dimensions we obtain a first general lower bound.

Proposition 3.1. $\kappa_{d,0} = \infty$ *for any* $d \geq 2$.

To overcome this problem, one can apply a continuous version of PCA taking into account (the dense set of) all points on the boundary of $CH(P)$, or even all points in $CH(P)$. In this approach X is a continuous set of d-dimensional vectors and the coefficients of the covariance matrix are defined by integrals instead of finite sums. If $CH(P)$ is known, the computation of the coefficients of the covariance matrix in the continuous case can also be done in linear time, thus, the overall complexity remains the same as in the discrete case. Note that for for $d = 1$ the above problem is trivial, because the PCA bounding box is always optimal, i.e., $\kappa_{1,0}$ is 1.

3.2 Continuous PCA

Variants of the continuous PCA applied to triangulated surfaces of 3D objects were presented by Gottschalk et. al. [14], Lahanas et. al. [25] and Vranić et. al. [50]. In what follows, we briefly review the basics of the continuous PCA in a general setting.

Let X be a continuous set of d-dimensional vectors with constant density. Then, the center of gravity of X is

$$c = \frac{\int_{x \in X} x \, dx}{\int_{x \in X} dx}. \qquad (3.1)$$

Here, $\int dx$ denotes either a line integral, an area integral, or a volume integral in higher dimensions. For any unit vector $v \in \mathbb{R}^d$, the *variance of X in direction v* is

$$var(X, v) = \frac{\int_{x \in X} \langle x - c, v \rangle^2 dx}{\int_{x \in X} dx}. \qquad (3.2)$$

The covariance matrix of X has the form

$$\Sigma = \frac{\int_{x \in X} (x - c)(x - c)^T dx}{\int_{x \in X} dx}, \qquad (3.3)$$

with its (i, j)-th component

$$\sigma_{ij} = \frac{\int_{x \in X} (x_i - c_i)(x_j - c_j) dx}{\int_{x \in X} dx}, \qquad (3.4)$$

where x_i and x_j are the i-th and j-th component of the vector x, and c_i and c_j the i-th and j-th component of the center of gravity. In the case when X is a continuous set of vectors, the procedure of finding the most significant directions can be also reformulated as an eigenvalue problem, and it can be verified that the results presented in Section 2.3.3 hold.

For point sets P in $\mathbb{R}^2$ we are especially interested in the cases when X represents the boundary of $CH(P)$, or all points in $CH(P)$. Since the first case corresponds to the 1-dimensional faces of $CH(P)$ and the second case to the only 2-dimensional face of $CH(P)$, the generalization to a dimension $d > 2$ leads to a series of $d-1$ continuous PCA versions. For a point set $P \in \mathbb{R}^d$, $\Sigma(P,i)$ denotes the covariance matrix defined by the points on the i-dimensional faces of $CH(P)$, and $BB_{pca(d,i)}(P)$, denotes the corresponding bounding box. The approximation factors $\kappa_{d,i}(P)$ and $\kappa_{d,i}$ are defined as

$$\kappa_{d,i}(P) = \frac{Vol(BB_{pca(d,i)}(P))}{Vol(BB_{opt}(P))}, \quad \text{and}$$
$$\kappa_{d,i} = \sup\left\{\kappa_{d,i}(P) \mid P \subseteq \mathbb{R}^d, Vol(CH(P)) > 0\right\}.$$

3.3 Lower Bounds

The lower bounds we are going to derive are based on the following connection between the symmetry of a point set and its principal components.

Lemma 3.1. *Let P be a d-dimensional point set symmetric with respect to a hyperplane H and assume that the covariance matrix Σ of P has d different eigenvalues. Then, a principal component of P is orthogonal to H.*

Proof. Without loss of generality, we can assume that the hyperplane of symmetry is spanned by the last $d-1$ standard base vectors of the d-dimensional space and the center of gravity of the point set coincides with the origin of the d-dimensional space, i.e., $c = (0, 0, \ldots, 0)$. Thus, we can write $P = P^+ \cup P^-$, where each point p^- from P^- has a

3.3. LOWER BOUNDS

counterpoint p^+ in P^+ (and vice versa) such that p^- and p^+ differ only in the first coordinate, namely $p_1^- = -p_1^+$. Then, we can rewrite (3.4) as

$$\sigma_{ij} = \frac{\int_{p \in P}(p_i - c_i)(p_j - c_j)dp}{\int_{p \in P} dp} = \frac{\int_{p \in P^+} p_i p_j dp}{\int_{p \in P^+} dp} + \frac{\int_{p \in P^-} p_i p_j dp}{\int_{p \in P^-} dp},$$

and

$$\sigma_{1j} = \frac{\int_{p \in P^+} p_1 p_j dp}{\int_{p \in P^+} dp} + \frac{\int_{p \in P^-} p_1 p_j dp}{\int_{p \in P^-} dp} = \frac{\int_{p \in P^+} p_1 p_j dp}{\int_{p \in P^+} dp} + \frac{\int_{p \in P^+} -p_1 p_j dp}{\int_{p \in P^+} dp}.$$

Then, the components σ_{1j}, for $2 \leq j \leq d$, are 0. Due to symmetry the components σ_{j1} are also 0. Thus, the covariance matrix has the form

$$\Sigma = \begin{pmatrix} \sigma_{11} & 0 & \cdots & 0 \\ 0 & \sigma_{22} & \cdots & \sigma_{2d} \\ \vdots & \vdots & \ddots & \vdots \\ 0 & \sigma_{d2} & \cdots & \sigma_{dd} \end{pmatrix}. \tag{3.5}$$

We note that the same argument carry thorough in the case when P is a discrete point set.

The characteristic polynomial of Σ is

$$\det(\Sigma - \lambda I) = (\sigma_{11} - \lambda)f(\lambda), \tag{3.6}$$

where $f(\lambda)$ is a polynomial of degree $d - 1$, with coefficients determined by the elements of the $(d-1) \times (d-1)$ submatrix of Σ. From this it follows that σ_{11} is a solution of the characteristic equation, i.e., it is an eigenvalue of Σ and the vector $(1, 0, ...,0)$ is its corresponding eigenvector (principal component), which is orthogonal to the assumed hyperplane of symmetry. □

We start with a generalization of Proposition 3.1.

Proposition 3.2. $\kappa_{d,i} = \infty$ for any $d \geq 4$ and any $1 \leq i < d - 1$.

Proof. We use a lifting argument to show that for any point set $P \subseteq \mathbb{R}^k$ there is a point set $P' \subseteq \mathbb{R}^{k+1}$ such that $\kappa_{k,i}(P) \leq \kappa_{k+1,i+1}(P')$, and consequently $\kappa_{k,i} \leq \kappa_{k+1,i+1}$.

Let Σ be the covariance matrix of P with eigenvalues $\lambda_1 > \lambda_2 > \cdots > \lambda_k$, and corresponding eigenvectors $v_1, v_2, \ldots v_k$. We define the point set $P'(h) = P \times [-h, h], h \in \mathbb{R}^+$. Let $\Sigma'(h)$ be the covariance matrix of $P'(h)$. Obviously, the point set $P'(h)$ is symmetric with respect to the hyperplane $H = \mathbb{R}^k \times \{0\}$, and by Lemma 3.1, the vector $v_{k+1} = (0, \ldots, 0, 1)$ is an eigenvector of $\Sigma'(h)$. Let $\lambda(h)$ be the corresponding eigenvalue of v_{k+1}. Since $\lambda(h) = \mathrm{var}(P', v_{k+1})$ is a quadratic function of h, with $\lim_{h \to 0} \lambda(h) = 0$, we can choose a value h_0 such that $\lambda(h_0)$ is smaller than the other eigenvalues of Σ'. Let v be an arbitrary direction in $\mathbb{R}^k$. Then, by definition of P', the variance of P' in the direction $(v, 0)$ remains the same as the variance of P in the direction v. Thus, we can conclude that the eigenvalues of Σ' are $\lambda_1 > \lambda_2 > \cdots > \lambda_k > \lambda(h_0)$, with corresponding eigenvectors $(v_1, 0), (v_2, 0), \ldots (v_k, 0), v_{k+1}$, and consequently $\mathrm{Vol}(BB_{pca(k+1,i+1)}(P')) = 2 h_0 \mathrm{Vol}(BB_{pca(k,i)}(P))$.

On the other hand, the bounding box $BB_{h_0} = BB_{opt}(P) \times [-h_0, h_0]$ is also a bounding box of P'. Therefore, we obtain

$$\kappa_{k+1,i+1} \geq \kappa_{k+1,i+1}(P') = \frac{\mathrm{Vol}(BB_{pca(k+1,i+1)}(P'))}{\mathrm{Vol}(BB_{opt}(P'))}$$
$$\geq \frac{\mathrm{Vol}(BB_{pca(k+1,i+1)}(P'))}{\mathrm{Vol}(BB_{h_0})} \geq \frac{2h_0 \mathrm{Vol}(BB_{pca(k,i)}(P))}{2h_0 \mathrm{Vol}(BB_{opt}(P))} \geq \kappa_{k,i}.$$

Now, we can establish $\kappa_{d,i} \geq \kappa_{d-1,i-1} \geq \ldots \geq \kappa_{d-i,0} = \infty$. $\square$

This way, there remain only two interesting cases for a given d: the factor $\kappa_{d,d-1}$ corresponding to the boundary of the convex hull, and the factor $\kappa_{d,d}$ corresponding to the full convex hull.

3.3.1 Lower bounds in $\mathbb{R}^2$

The result obtained in this subsection can be seen as a special case of the result obtained in Subsection 3.3.3. To gain a better understanding of the problem and the obtained results, we consider it separately.

Theorem 3.1. $\kappa_{2,1} \geq 2$ and $\kappa_{2,2} \geq 2$.

Proof. Both lower bounds can be derived from a rhombus. Let the side length of the rhombus be 1. To make sure that the covariance matrix has two distinct eigenvalues, we assume that the rhombus has an angle $\alpha > 90°$. Since the rhombus is symmetric, its PCs coincide with its diagonals. In Figure 3.2 (b) its optimal-area bounding boxes, for 2 different angles, $\alpha > 90°$ and $\beta = 90°$, are shown, and in Figure 3.2 (a) its corresponding PCA bounding boxes. When the rhombus' angles in limit approach $90°$, the rhombus approaches a square with side length 1, i.e., the vertices of the rhombus in the limit are $(\frac{1}{\sqrt{2}}, 0), (-\frac{1}{\sqrt{2}}, 0), (0, \frac{1}{\sqrt{2}})$ and $(0, -\frac{1}{\sqrt{2}})$ (see Figure 3.2 (a)), and the area of its PCA bounding box is $\sqrt{2} \times \sqrt{2}$. According to Lemma 3.1, the PCs of the rhombus are unique as long its angles are not $90°$. This leads to the conclusion that the ratio between the area of the PCA bounding box in Figure 3.2 (a) and the area of the optimal-area bounding box in Figure 3.2 (b) in limit goes to 2. □

Alternatively, to show that the given squared rhombus fits into a unit square, one can apply the following rotation matrix

$$R_2 = \frac{1}{\sqrt{2}} \begin{pmatrix} 1 & 1 \\ 1 & -1 \end{pmatrix}. \tag{3.7}$$

It can be verified easily that all coordinates of the vertices of the rhombus transformed by R_2 are in the interval $[-0.5, 0.5]$. We use similar arguments when we prove the lower bounds in higher dimensions.

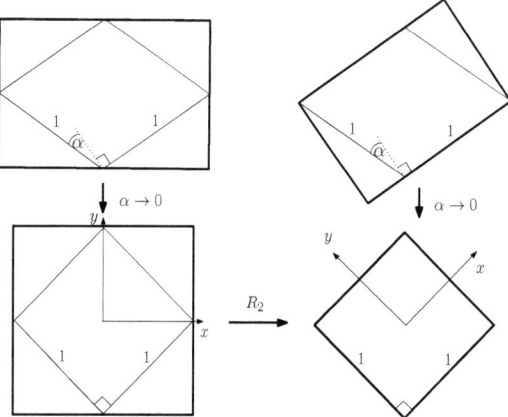

Figure 3.2: An example which gives the lower bound on the area of the PCA bounding box of an arbitrary convex polygon in $\mathbb{R}^2$.

3.3.2 Lower bounds in $\mathbb{R}^3$

Theorem 3.2. $\kappa_{3,2} \geq 4$ and $\kappa_{3,3} \geq 4$.

Proof. Both lower bounds are obtained from a dipyramid, having a rhombus with side length $\sqrt{2}$ as its base. The other sides of the dipyramid have length $\frac{\sqrt{3}}{2}$. Similarly as in $\mathbb{R}^2$, we consider the case when its base, the rhombus, in limit approaches the square, i.e., the vertices of the square dipyramid are $(1, 0, 0), (-1, 0, 0), (0, 1, 0), (0, -1, 0)$, $(0, 0, \frac{\sqrt{2}}{2})$ and $(0, 0, -\frac{\sqrt{2}}{2})$ (see Figure 3.3 (a)). The side lengths of its PCA bounding box are 2, 2 and $\sqrt{2}$. Now, we rotate the coordinate system (or the square dipyramid) with the rotation determined by the

3.3. LOWER BOUNDS

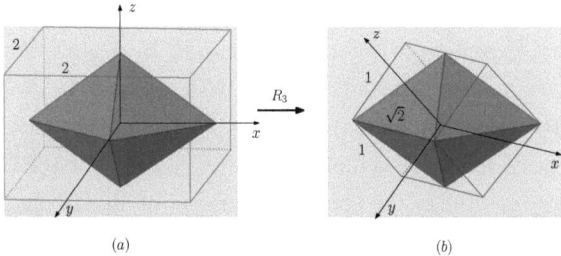

Figure 3.3: An example which gives the lower bound on the volume of the PCA bounding box of an arbitrary convex polygon in $\mathbb{R}^3$.

following orthogonal matrix

$$R_3 = \begin{pmatrix} \frac{1}{\sqrt{2}} & -\frac{1}{\sqrt{2}} & 0 \\ \frac{1}{2} & \frac{1}{2} & -\frac{1}{\sqrt{2}} \\ \frac{1}{2} & \frac{1}{2} & \frac{1}{\sqrt{2}} \end{pmatrix}. \tag{3.8}$$

It can be verified easily that the square dipyramid, after rotation with R_3 fits into the cube $[-0.5, 0.5]^3$ (see Figure 3.3 (b)). Thus, the ratio of the volume of the bounding box, Figure 3.3 (a), and the volume of its PCA bounding box, Figure 3.3 (b), in limit goes to 4. □

3.3.3 Lower bounds in $\mathbb{R}^d$

The lower bounds, presented in this subsection, are based on the following result.

Theorem 3.3. *If the dimension d of the bounding box is*

(a) a power of two, or

(b) a multiply of four and at most 664,

then $\kappa_{d,d-1} \geq d^{d/2}$ and $\kappa_{d,d} \geq d^{d/2}$.

Proof. (a) For any $d = 2^k$, $k \in \mathbb{N}\backslash\{0\}$, let $\mathbf{a}_i$ be a d-dimensional vector, with $a_{ii} = \frac{\sqrt{d}}{2}$ and $a_{ij} = 0$ for $i \neq j$, and let $\mathbf{b}_i = -\mathbf{a}_i$. We construct a d-dimensional convex polytope P_d with vertices $V = \{\mathbf{a}_i, \mathbf{b}_i | 1 \leq i \leq d\}$. It is easy to check that the hyperplane orthogonal to $\mathbf{a}_i$ is a hyperplane of reflective symmetry, and as consequence of Lemma 3.1, $\mathbf{a}_i$ is an eigenvector of the covariance matrix of P_d. To ensure that all eigenvalues are different (which implies that the PCA bounding box is unique), we add $\epsilon_i > 0$ to the i-th coordinate of $\mathbf{a}_i$, and $-\epsilon_i$ to the i-th coordinate of $\mathbf{b}_i$, for $1 \leq i \leq d$, where $\epsilon_1 < \epsilon_2 < \cdots < \epsilon_d$. When all ϵ_i, $1 \leq i \leq d$, tend to 0, the PCA bounding box of the convex polytope P_d converges to a hypercube with side lengths $\sqrt{d}$, i.e., the volume of the PCA bounding box of P_d converges to $d^{d/2}$. Now, we rotate P_d, such that it fits into the cube $[-\frac{1}{2}, \frac{1}{2}]^d$. For $d = 2^k$, we can use a rotation matrix

$$R_d = \frac{1}{\sqrt{2}} \begin{pmatrix} R_{\frac{d}{2}} & R_{\frac{d}{2}} \\ \hline R_{\frac{d}{2}} & -R_{\frac{d}{2}} \end{pmatrix}, \qquad (3.9)$$

where we start with the matrix $R_1 = (1)$. A straightforward calculation verifies that P_d rotated with R_d fits into the cube $[-0.5, 0.5]^d$.

(b) Before we prove this part of the theorem, we would like to note that the derivation of R_d in (a) can be traced back to a Hadamard matrix.

A *Hadamard matrix* of order $d \times d$, denoted by H_d, is a ± 1 matrix with orthogonal columns.

Alternatively, we can define R_d as

$$R_d = \frac{1}{\sqrt{d}} H_d, \qquad (3.10)$$

where

$$H_d = \begin{pmatrix} H_{\frac{d}{2}} & H_{\frac{d}{2}} \\ \hline H_{\frac{d}{2}} & -H_{\frac{d}{2}} \end{pmatrix}, \qquad (3.11)$$

3.3. LOWER BOUNDS

and
$$H_2 = \begin{pmatrix} 1 & 1 \\ 1 & -1 \end{pmatrix}. \tag{3.12}$$

From the construction in the proof for (a), it follows that the theorem holds for all dimensions d for which a $d \times d$ Hadamard matrix exists. In (a), it was shown that a Hadamard matrix always exits when $d = 2^k$, $k \in \mathbb{N} \setminus \{0\}$. Hadamard conjectured that a Hadamard matrix also exists when $d = 4k$, $k \in \mathbb{N} \setminus \{0\}$. This conjecture is known to be true for $d \leq 664$ [23]. □

We can combine lower bounds from lower dimensions to get lower bounds in higher dimensions by taking Cartesian products. If κ_{d_1} is a lower bound on the ratio between the PCA bounding box and the optimal bounding box of a convex polytope in $\mathbb{R}^{d_1}$, and κ_{d_2} is a lower bound in $\mathbb{R}^{d_2}$, then $\kappa_{d_1} \cdot \kappa_{d_2}$ is a lower bound in $\mathbb{R}^{d_1+d_2}$. This observation together with the results from this section enables us to obtain lower bounds in any dimension. For example, for the first 12

Table 3.1: Lower bounds for the approximation factor of PCA bounding boxes for the first 12 dimensions.

dim.	$\mathbb{R}$	$\mathbb{R}^2$	$\mathbb{R}^3$	$\mathbb{R}^4$	$\mathbb{R}^5$	$\mathbb{R}^6$	$\mathbb{R}^7$	$\mathbb{R}^8$	$\mathbb{R}^9$	$\mathbb{R}^{10}$	$\mathbb{R}^{11}$	$\mathbb{R}^{12}$
lower bound	1	2	4	16	16	32	64	4096	4096	8192	16384	2985984

dimensions, the lower bounds we obtain are given in Table 3.1.

One can observe big gaps between the bounds in $\mathbb{R}^7$ and $\mathbb{R}^8$, and between the bounds in $\mathbb{R}^{11}$ and $\mathbb{R}^{12}$. The bound in $\mathbb{R}^7$ is obtained as a product of the lower bounds in $\mathbb{R}^3$ and $\mathbb{R}^4$, and the bound in $\mathbb{R}^{11}$ is obtained as a product of the lower bounds in $\mathbb{R}^3$ and $\mathbb{R}^8$, while the bounds in $\mathbb{R}^8$ and $\mathbb{R}^{12}$ are obtained directly from Theorem 3.3. This indicates that for dimensions that are not covered by Theorem 3.3, one can expect much bigger lower bounds. It is an interesting open problem to develop techniques for such improvements.

Chapter 4

Upper Bounds on the Quality of PCA Bounding Boxes

In this chapter, we present upper bounds on the approximation factors of PCA bounding boxes in $\mathbb{R}^2$ and $\mathbb{R}^3$. As it was shown in Proposition 3.1, the considered bounds for discrete point sets tend to infinity. Thus, we are interested in PCA bounding boxes for continuous point sets, especially for the convex hull of point sets. In Proposition 3.2, it was shown that the only two cases, related to the convex hull of the point set, when the approximation factor does not tend to infinity, are those when the whole convex hull, or the boundary of the convex hull are considered. The corresponding approximation factors were denoted by $\kappa_{d,d}$ and $\kappa_{d,d-1}$. In this chapter, we present upper bounds on $\kappa_{2,1}$, $\kappa_{2,2}$ and $\kappa_{3,3}$.

Starting from the principle that the study of the worst case examples (established by the known lower bounds) could give an idea how to prove upper bounds, we make a surprising observation: Since most of the worst case examples have minimum-volume bounding boxes with unit lengths of all sides, it is trivial that any bounding box approximates with a factor at most $\sqrt{d}^d$. Thus, we have a trivial upper bound for all point sets with an optimal bounding box of unit lengths of all sides. Moreover, in $\mathbb{R}^2$ this argument can be generalized to a parameterized upper bound depending on the ratio η between the lengths of the longest and the shortest side of the minimum-volume bounding box. Again, this is not a special upper bound for the PCA-algorithm, it applies to all "tight" bounding boxes with respect to any orthonormal coordinate system. Thus, we need a second upper bound argument that makes use of the special properties of PCA, and that works well when the ratio η is large. To obtain the final upper bound we consider the lower envelop of both parameterized bounds and search for its maximum (over all $\eta \geq 1$).

Upper bounds in $\mathbb{R}^2$. The first parameterized bound in $\mathbb{R}^2$ is common for both $\kappa_{2,1}$ and $\kappa_{2,2}$. It depends on the parameter η. The bound is presented in Lemma 4.1, and it is based on a simple estimation of the diameter of the point set. It is good for a small values of the parameter η. An improvement of this bound is presented in Lemma 4.5 and it is obtained by computing the maximum area rectangle that touches a certain rectangle.

The second parameterized bounds on $\kappa_{2,1}$ and $\kappa_{2,2}$ are presented in Lemmas 4.4 and 4.9, respectively. Both are good for big values of the parameter η. The essence of deriving these bounds is an estimation of the distance of the continuous point set to its best fitting line. However, the techniques used to obtain the estimations differ for $\kappa_{2,1}$ and for $\kappa_{2,2}$. For $\kappa_{2,1}$, we exploit arguments from discrete geometry (Lemmas 4.2 and 4.3), while for $\kappa_{2,2}$ we use ideas from integral calculus (Theorems 4.3 and 4.4).

An upper bound in $\mathbb{R}^3$. We present an upper bound on $\kappa_{3,3}$. We follow the ideas from the derivation of the upper bounds on $\kappa_{2,2}$. However, in $\mathbb{R}^3$ there are two main differences with respect to $\mathbb{R}^2$. First, the density function of the convex hull in $\mathbb{R}^3$ (see Section 4.2.1 for the definition) is not convex. Instead of convexity, another property of the density function (described in Proposition 4.1) is used. Second,

4.1 Upper Bounds in $\mathbb{R}^2$

4.1.1 An upper bound on $\kappa_{2,1}$

Given a point set $P \subseteq \mathbb{R}^2$ and an arbitrary bounding box $BB(P)$ we will denote the two side lengths by a and b, where $a \geq b$. We are interested in the side lengths $a_{opt}(P) \geq b_{opt}(P)$ and $a_{pca}(P) \geq b_{pca}(P)$ of $BB_{opt}(P)$ and $BB_{pca(2,1)}(P)$, see Figure 4.1. The parameters $\alpha = \alpha(P) = a_{pca}(P)/a_{opt}(P)$ and $\beta = \beta(P) = b_{pca}(P)/b_{opt}(P)$ denote the ratios between the corresponding side lengths. Hence, we have $\kappa_{2,1}(P) = \alpha(P) \cdot \beta(P)$. If the relation to P is clear, we will omit the reference to P in the notations introduced above.

Since the side lengths of any bounding box are bounded by the diameter of P, we can observe that in general $b_{pca}(P) \leq a_{pca}(P) \leq diam(P) \leq \sqrt{2} a_{opt}(P)$, and in the special case when the optimal bounding box is a square $\kappa_{2,1}(P) \leq 2$. This observation can be generalized, introducing an additional parameter $\eta(P) = a_{opt}(P)/b_{opt}(P)$.

Lemma 4.1. $\kappa_{2,1}(P) \leq \eta + \frac{1}{\eta}$ and $\kappa_{2,2}(P) \leq \eta + \frac{1}{\eta}$ for any point set P with fixed aspect ratio $\eta(P) = \eta$.

Proof. We have for both a_{pca} and b_{pca} the upper bound $diam(P) \leq \sqrt{a_{opt}^2 + b_{opt}^2} = a_{opt}\sqrt{1 + \frac{1}{\eta^2}}$. Replacing a_{opt} by $\eta \cdot b_{opt}$ in the bound on b_{pca} we obtain $\alpha\beta \leq \eta \left(\sqrt{1 + \frac{1}{\eta^2}}\right)^2 = \eta + \frac{1}{\eta}$. □

Unfortunately, this parameterized upper bound tends to infinity for $\eta \to \infty$. Therefore, we are going to derive another upper bound that is better for large values of η. In this process we will make essential

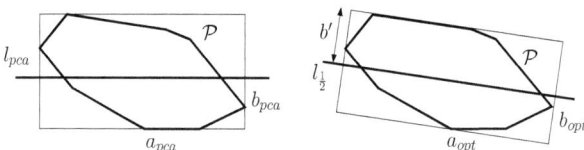

Figure 4.1: A convex polygon $\mathcal{P}$, its PCA bounding box and the line l_{pca}, which coincides with the first principal component of $\mathcal{P}$ (a). The optimal bounding box and the line $l_{\frac{1}{2}}$, going through the middle of its smaller side, parallel with its longer side (b).

use of the properties of $BB_{pca(2,1)}(P)$. In order to distinguish clearly between a convex set and its boundary, we will use calligraphic letters for the boundaries, specifically $\mathcal{P}$ for the boundary of $CH(P)$ and $\mathcal{BB}_{opt}$ for the boundary of the rectangle $BB_{opt}(P)$. Furthermore, we denote by $d^2(\mathcal{P}, l)$ the integral of the squared distances of the points on $\mathcal{P}$ to a line l, i.e., $d^2(\mathcal{P}, l) = \int_{x \in \mathcal{P}} d^2(x, l) ds$. Let l_{pca} be the line going through the center of gravity and parallel to the longer side of $BB_{pca(2,1)}(P)$ and $l_{\frac{1}{2}}$ be the bisector of $BB_{opt(P)}$ parallel to the longer side. By Proposition 2.5, l_{pca} is the best fitting line of P and therefore,

$$d^2(\mathcal{P}, l_{pca}) \leq d^2(\mathcal{P}, l_{\frac{1}{2}}). \tag{4.1}$$

Lemma 4.2. $d^2(\mathcal{P}, l_{\frac{1}{2}}) \leq \frac{b_{opt}^2 a_{opt}}{2} + \frac{b_{opt}^3}{6}$.

Proof. If a segment of $\mathcal{P}$ intersects the line $l_{\frac{1}{2}}$, we split this segment into two segments, with the intersection point as a split point. Then, to each segment f of $\mathcal{P}$ flush with the side of the PCA bounding box, we assign a segment identical to f. To each remaining segment s of $\mathcal{P}$, with endpoints (x_1, y_1) and (x_2, y_2), where $|y_1| \leq |y_2|$, we assign two segments: a segment s_1, with endpoints (x_1, y_1) and (x_1, y_2), and a segment s_2, with endpoints (x_1, y_2) and (x_2, y_2). All these segments form the boundary $\mathcal{BB}_S$ of a staircase polygon (see Figure 4.2 for illustration). Two straightforward consequences are that $d^2(\mathcal{BB}_S, l_{\frac{1}{2}}) \leq$

4.1. UPPER BOUNDS IN $\mathbb{R}^2$

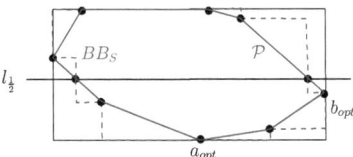

Figure 4.2: The convex polygon $\mathcal{P}$, its optimal bounding box, and the staircase polygon BB_S (depicted dashed).

$d^2(\mathcal{BB}_{opt}, l_{\frac{1}{2}})$, and $d^2(s, l_{\frac{1}{2}}) \leq d^2(s_1, l_{\frac{1}{2}}) + d^2(s_2, l_{\frac{1}{2}})$, for each segment s of $\mathcal{P}$. Therefore, $d^2(\mathcal{P}, l_{\frac{1}{2}})$ is at most $d^2(\mathcal{BB}_S, l_{\frac{1}{2}})$, which is bounded from above by $d^2(\mathcal{BB}_{opt}, l_{\frac{1}{2}}) = 4 \int_0^{\frac{b_{opt}}{2}} x^2 \, dx + 2 \int_0^{a_{opt}} (\frac{b_{opt}}{2})^2 \, dx = \frac{b_{opt}^2 a_{opt}}{2} + \frac{b_{opt}^3}{6}$. □

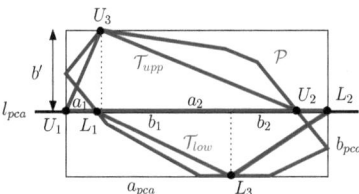

Figure 4.3: The convex polygon $\mathcal{P}$, its PCA bounding box, and a construction for a lower bound on $d^2(\mathcal{P}, l_{pca})$.

Now we look at $\mathcal{P}$ and its PCA bounding box (Figure 4.3). The line l_{pca} divides $\mathcal{P}$ into an upper and a lower part, $\mathcal{P}_{upp}$ and $\mathcal{P}_{low}$. l_{upp} denotes the orthogonal projection of $\mathcal{P}_{upp}$ onto l_{pca}, with U_1 and U_2 as its extreme points, and l_{low} denotes the orthogonal projection of $\mathcal{P}_{low}$ onto l_{pca}, with L_1 and L_2 as its extreme points. $\mathcal{T}_{upp} = \triangle(U_1 U_2 U_3)$ is a triangle inscribed in the upper part of the PCA bounding box (the part above l_{pca}), where point U_3 lies on the intersection of $\mathcal{P}_{upp}$ with the upper side of the PCA bounding box. Analogously, $\mathcal{T}_{low} = \triangle(L_1 L_2 L_3)$ is a triangle inscribed in the lower part of the PCA bounding box (the

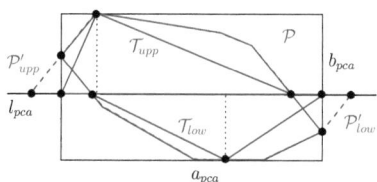

Figure 4.4: Two polylines $\mathcal{P}'_{upp}$ and $\mathcal{P}'_{low}$ (depicted dashed) formed from $\mathcal{P}$.

part below l_{pca}).

Lemma 4.3.

$$d^2(\mathcal{P}, l_{pca}) \geq d^2(\mathcal{T}_{upp}, l_{pca}) + d^2(\mathcal{T}_{low}, l_{pca}).$$

Proof. Let Q denote a chain of segments of $\mathcal{P}$, which does not touch the longer side of the PCA bounding box, and whose one endpoint lies on the smaller side of the PCA bounding box, and the other endpoint on the line l_{pca}. We reflect Q at the line supporting the side of the PCA bounding box touched by Q. All such reflected chains of segments, together with the rest of $\mathcal{P}$, form two polylines: $\mathcal{P}'_{upp}$ and $\mathcal{P}'_{low}$ (see Figure 4.4 for illustration). As a consequence, to each of the sides of the triangles $\mathcal{T}_{low}$ and $\mathcal{T}_{upp}$, $\overline{L_1L_3}$, $\overline{L_2L_3}$, $\overline{U_1U_3}$, $\overline{U_2U_3}$, we have a corresponding chain of segments R as shown in the two cases in Figure 4.5. In both cases $d^2(t, l_{pca}) \leq d^2(R, l_{pca})$. Namely, we can parametrize both curves, R and t, starting at the common endpoint A that is furthest from l_{pca}. By comparing two points with the same parameter (distance from A along the curve) we see that the point on t always has a smaller distance to l_{pca} than the corresponding point on R. In addition t is shorter, and some parts of R have no match on t.

Consequently, $d^2(\mathcal{P}', l_{pca}) \geq d^2(\mathcal{T}_{upp} \bigcup \mathcal{T}_{low}, l_{pca}) = d^2(\mathcal{T}_{upp}, l_{pca}) + d^2(\mathcal{T}_{low}, l_{pca})$, and since $d^2(\mathcal{P}', l_{pca}) = d^2(\mathcal{P}, l_{pca}) = d^2(\mathcal{P}_{upp} \bigcup \mathcal{P}_{low}, l_{pca})$, the proof is completed. □

4.1. UPPER BOUNDS IN $\mathbb{R}^2$

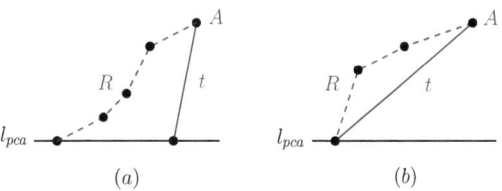

Figure 4.5: Two types of chains of segments (depicted dashed and denoted by R), and their corresponding triangles' edges (depicted solid and denoted by t). The base-point of t corresponds to the most left point of $\mathcal{T}_{upp}$ from Figure 4.3 and Figure 4.4.

Since $\mathcal{P}$ is convex, the following relations hold:

$$|l_{upp}| \geq \frac{b'}{b_{pca}} a_{pca}, \text{ and } |l_{low}| \geq \frac{b_{pca} - b'}{b_{pca}} a_{pca}. \tag{4.2}$$

The value

$$\begin{aligned} d^2(\mathcal{T}_{upp}, l_{pca}) &= \int_0^{\sqrt{a_1^2+b'^2}} (\frac{\alpha}{\sqrt{a_1^2+b'^2}} b')^2 \, d\alpha + \int_0^{\sqrt{a_2^2+b'^2}} (\frac{\alpha}{\sqrt{a_2^2+b'^2}} b')^2 \, d\alpha \\ &= \frac{b'^2}{3}(\sqrt{a_1^2+b'^2} + \sqrt{a_2^2+b'^2}) \end{aligned}$$

is minimal when $a_1 = a_2 = \frac{|l_{upp}|}{2}$. With (4.2) we get

$$d^2(\mathcal{T}_{upp}, l_{pca}) \geq \frac{b'^3}{3 b_{pca}} \sqrt{a_{pca}^2 + 4 b_{pca}^2}.$$

Analogously, we have for the lower part:

$$d^2(\mathcal{T}_{low}, l_{pca}) \geq \frac{(b_{pca} - b')^3}{3 b_{pca}} \sqrt{a_{pca}^2 + 4 b_{pca}^2}.$$

The sum $d^2(\mathcal{T}_{upp}, l_{pca}) + d^2(\mathcal{T}_{low}, l_{pca})$ is minimal when $b' = \frac{b_{pca}}{2}$. This, together with Lemma 4.3, gives:

$$d^2(\mathcal{P}, l_{pca}) \geq \frac{b_{pca}^2}{12} \sqrt{a_{pca}^2 + 4 b_{pca}^2}. \tag{4.3}$$

Combining (4.1), (4.3) and Lemma 4.2 we have:

$$\frac{1}{2}a_{opt}b_{opt}^2 + \frac{1}{6}b_{opt}^3 \geq \frac{b_{pca}^2}{12}\sqrt{a_{pca}^2 + 4b_{pca}^2} \geq \frac{b_{pca}^2}{12}a_{pca}. \quad (4.4)$$

Replacing a_{opt} with ηb_{opt} on the left side, b_{pca}^2 with $\beta^2 b_{opt}^2$ and a_{pca} with $\alpha a_{opt} = \alpha \eta b_{opt}$ on the right side of (4.4), we obtain:

$$\left(\frac{\eta}{2} + \frac{1}{6}\right)b_{opt}^3 \geq \frac{\beta^2 \alpha \eta}{12}b_{opt}^3$$

which implies

$$\beta \leq \sqrt{\frac{6\eta + 2}{\alpha \eta}}.$$

This gives the second upper bound on $\kappa_{2,1}(P)$ for point sets with parameter η:

$$\alpha\beta \leq \sqrt{\frac{(6\eta + 2)\alpha}{\eta}} \leq \sqrt{\frac{6\eta + 2}{\eta}}\sqrt{1 + \frac{1}{\eta^2}}. \quad (4.5)$$

Lemma 4.4. $\kappa_{2,1}(P) \leq \sqrt{\frac{6\eta+2}{\eta}}\sqrt{1 + \frac{1}{\eta^2}}$ for any point set P with fixed aspect ratio $\eta(P) = \eta$.

This implies the final result of this subsection.

Theorem 4.1. *The PCA bounding box of a point set P in $\mathbb{R}^2$ computed over the boundary of $CH(P)$ has a guaranteed approximation factor $\kappa_{2,1} \leq 2.737$.*

Proof. The theorem follows from the combination of the two parameterized bounds from Lemma 4.1 and Lemma 4.4 proved above:

$$\kappa_{2,1} \leq \sup_{\eta \geq 1}\left\{\min\left(\eta + \frac{1}{\eta}, \sqrt{\frac{6\eta + 2}{\eta}}\sqrt{1 + \frac{1}{\eta^2}}\right)\right\}.$$

It is easy to check that the supremum $s \approx 2.736$ is obtained for $\eta \approx 2.302$. □

4.1.2 An improved upper bound on $\kappa_{2,1}$

An improvement of the upper bound on $\kappa_{2,1}$ is obtained by providing better upper bound on $\kappa_{2,1}$ for big η than that from Lemma 4.1.

Given two rectangles R_1 and R_2 in $\mathbb{R}^2$, we say that R_2 *touches* R_1 if the intersection of each side of R_2 with R_1 is a corner of R_1, or a side of R_1. Note that in the latter case, R_1 and R_2 are identical.

Lemma 4.5. *Let R_1 be a rectangle in $\mathbb{R}^2$, with side's lengths a and b, such that $a = \eta b$, $\eta \geq 1$. Then, the area of the maximum-area rectangle that touches R_1 is $\frac{(\eta+1)^2 b^2}{2}$.*

Proof. Let R_2 denote the maximum-area rectangle that touches R_1. Due to Thales theorem, the corners of R_2 must lie on the half-circles built above each side of R_1. See Figure 4.6 for an illustration. The

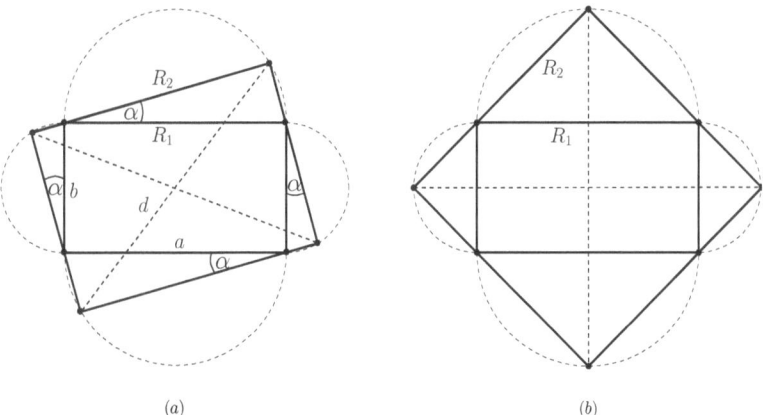

Figure 4.6: *(a)* A rectangle R_1 and a touching rectangle R_2 of R_1. *(b)* The maximum-area touching rectangle R_2 of R_1.

area of R_2 equals to the area of R_1 plus the area of the four similar triangles (the triangles with angle α in Figure. 4.6). Then

$$\begin{aligned} \text{area}(R_2) &= ab + a^2 \sin\alpha\cos\alpha + b^2 \sin\alpha\cos\alpha \\ &= ab + (a^2+b^2)\frac{\sin 2\alpha}{2}. \end{aligned}$$

This expression has maximal value for $\alpha = \frac{\pi}{4}$, namely $\frac{(a+b)^2}{2}$. Since $a = \eta b$, we finally have that the maximal area of R_2 is $\frac{(\eta+1)^2 b^2}{2}$. □

Lemma 4.6. $\kappa_{2,1}(P) \leq \frac{(\eta+1)^2}{2\eta}$ and $\kappa_{2,2}(P) \leq \frac{(\eta+1)^2}{2\eta}$ for any point set P with fixed aspect ratio $\eta(P) = \eta$.

Proof. For a point set P with fixed aspect ratio η we have that $\text{area}(BB_{opt}(P)) = a_{opt} b_{opt} = \eta b_{opt}^2$. Denote by BB_{max} the maximum-area rectangle that touches BB_{opt}. By Lemma 4.5 it follows that $\text{area}(BB_{max}) = \frac{(\eta+1)^2 b_{opt}^2}{2}$. Thus, we have

$$\frac{\text{area}(BB_{pca}(P))}{\text{area}(BB_{opt}(P))} \leq \frac{\text{area}(BB_{max}(P))}{\text{area}(BB_{opt}(P))} = \frac{(\eta+1)^2 b_{opt}^2}{2\eta b_{opt}^2} = \frac{(\eta+1)^2}{2\eta}.$$

□

Theorem 4.2. *The PCA bounding box of a point set P in $\mathbb{R}^2$ computed over $CH(P)$ has a guaranteed approximation factor $\kappa_{2,1} \leq 2.654$.*

Proof. The theorem follows from the combination of the two parameterized bounds from Lemma 4.4 and Lemma 4.6:

$$\kappa_{2,1} \leq \sup_{\eta \geq 1} \left\{ \min\left(\frac{(\eta+1)^2}{2\eta}, \sqrt{\frac{6\eta+2}{\eta}\sqrt{1+\frac{1}{\eta^2}}} \right) \right\}.$$

The supremum for the expression above occurs at 2.6535 for $\eta \approx 2.970$. □

Although these results concern a continuous PCA version, the proofs are mainly based on arguments from discrete geometry. In contrast

4.1. UPPER BOUNDS IN $\mathbb{R}^2$

to that, the upper bound proofs for $\kappa_{2,2}$ and $\kappa_{3,3}$, presented in the next two subsections, essentially make use of integral calculus.

4.1.3 An upper bound on $\kappa_{2,2}$

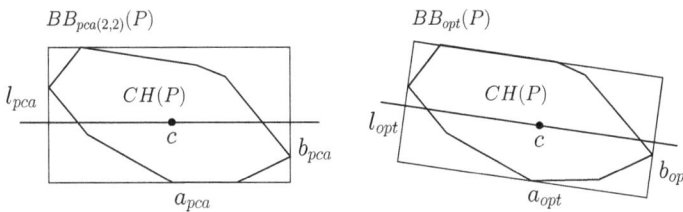

Figure 4.7: A convex hull of the point set P, its PCA bounding box (a) and its optimal bounding box (b).

First, we note that due to Lemma 4.1, we already have a parameterized upper bound on $\kappa_{2,2}$. Since this bound tends to infinity for $\eta \to \infty$, we are going to derive another upper bound on $\kappa_{2,2}$ that is better for large values of η. We derive such a bound by finding a constant that bounds β from above. In this process we will make essential use of the properties of $BB_{pca(2,2)}(P)$. We denote by $d^2(CH(P), l)$ the integral of the squared distances of the points on $CH(P)$ to a line l, i.e.,

$$d^2(CH(P), l) = \int_{s \in CH(P)} d^2(s, l) ds.$$

Let l_{pca} be the line going through the center of gravity, parallel to the longer side of $BB_{pca(2,2)}(P)$, and l_{opt} be the line going through the center of gravity, parallel to the longer side of $BB_{opt(P)}$ (see Figure 4.7). By Proposition 2.5, l_{pca} is the best fitting line of P and therefore,

$$d^2(CH(P), l_{pca}) \leq d^2(CH(P), l_{opt}). \qquad (4.6)$$

We obtain an estimate of β by determining a lower bound on $d^2(CH(P), l_{pca})$ that depends on b_{pca}, and an upper bound on $d^2(CH(P), l_{opt})$ that depends on b_{opt}. Having an arbitrary bounding box of $CH(P)$ (with side lengths a and b, $a \geq b$) the area of $CH(P)$ can be expressed as

$$A = A(CH(P)) = \int_0^b \int_0^a \chi_{CH(P)}(x,y) dx dy = \int_0^b g(y) dy,$$

where $\chi_{CH(P)}(x,y)$ is the *characteristic function* of $CH(P)$ defined as

$$\chi_{CH(P)}(x,y) = \begin{cases} 1 & (x,y) \in CH(P) \\ 0 & (x,y) \notin CH(P), \end{cases}$$

and $g(y) = \int_0^a \chi_{CH(P)}(x,y) dx$ is the length of the intersection of $CH(P)$ with a horizontal line at height y. In the following we call $g(y)$ the *density function* of $CH(P)$ for computing the area with the integral $\int_0^b g(y) dy$. Note that $g(y)$ is continuous and convex in the interval $[0, b]$ (see Figure 4.8 for an illustration). Let b_1 denote the

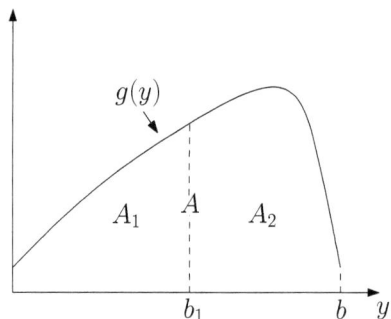

Figure 4.8: Density function $g(y)$ of the convex hull of a point set in $\mathbb{R}^2$.

y-coordinate of the center of gravity of $CH(P)$. The line l_{b_1} ($y = b_1$) divides the area of $CH(P)$ into A_1 and A_2.

Theorem 4.4, which is derived from the generalized first mean value theorem of integral calculus (Theorem 4.3), is our central technical tool in derivation of the lower and the upper bound on $d^2(CH(P), l_{b_1})$.

4.1. UPPER BOUNDS IN $\mathbb{R}^2$

Theorem 4.3. (Generalized first mean value theorem of integral calculus)

If $h(x)$ and $g(x)$ are continuous functions on the interval $[a,b]$, and if $g(x)$ does not change its sign in the interval, then there is a $\xi \in (a,b)$ such that

$$\int_a^b h(x)g(x)dx = h(\xi)\int_a^b g(x)dx.$$

Theorem 4.4. Let $f(x)$ and $g(x)$ be positive continuous functions on the interval $[a,b]$ with $\int_a^b f(x)dx = \int_a^b g(x)dx$, and assume that there is some $c \in [a,b]$ such that $f(x) \le g(x)$, for all $x \le c$ and $f(x) \ge g(x)$, for all $x \ge c$. Then

$$\int_a^b (x-b)^2 f(x)dx \le \int_a^b (x-b)^2 g(x)dx \quad \text{and}$$

$$\int_a^b (x-a)^2 f(x)dx \ge \int_a^b (x-a)^2 g(x)dx.$$

Proof. We start from the assumptions $\int_a^b f(x)dx = \int_a^b g(x)dx$ and $f(x) \le g(x)$ for all $x \le c$ and $f(x) \ge g(x)$ for all $x \ge c$. Thus,

$$\int_a^c (g(x) - f(x))dx = \int_c^b (f(x) - g(x))dx = \Delta \qquad (4.7)$$

and the integrands on both sides are nonnegative. Applying Theorem 4.3 to the following integrals we obtain

$$\int_a^c (x-b)^2 (g(x) - f(x))dx = (\xi_1 - b)^2 \int_a^c (g(x) - f(x))dx$$
$$= (\xi_1 - b)^2 \Delta,$$

and

$$\int_c^b (x-b)^2 (f(x) - g(x))dx = (\xi_2 - b)^2 \int_c^b (f(x) - g(x))dx$$
$$= (\xi_2 - b)^2 \Delta,$$

for some $\xi_1 \in [a, c]$ and $\xi_2 \in [c, b]$. Therefore,

$$\int_a^c (x-b)^2(g(x)-f(x))dx = (\xi_1 - b)^2 \Delta \geq (\xi_2 - b)^2 \Delta$$
$$= \int_c^b (x-b)^2(f(x)-g(x))dx.$$

It follows that

$$\int_a^b (x-b)^2(g(x)-f(x))dx = \int_a^c (x-b)^2(g(x)-f(x))dx -$$
$$\int_c^b (x-b)^2(f(x)-g(x))dx$$
$$\geq 0,$$

which proves the first claim

$$\int_a^b (x-b)^2 f(x)dx \leq \int_a^b (x-b)^2 g(x)dx.$$

The proof of the second claim follows by symmetry. □

The following theorem was discovered independently by Grünbaum [15] and Hammer (unpublished manuscript), and later rediscovered by Mityagin [31]. We use it to prove a lower and an upper bound of the variance $d^2(CH(P), l_{b_1})$.

Theorem 4.5 (Grünbaum-Hammer-Mityagin). *Let K be a compact convex set in $\mathbb{R}^d$ with non-empty interior and centroid μ. Assume that the d-dimensional volume of K is one, that is, $Vol_d(K) = 1$. Let H be any (d-1)-dimensional hyperplane passing through μ with corresponding half-spaces H^+ and H^-. Then,*

$$\min\{Vol_d(K \cap H^+), Vol_d(K \cap H^-)\} \geq \left(\frac{d}{d+1}\right)^d.$$

Moreover, the bound $(\frac{d}{d+1})^d$ is best possible.

4.1. UPPER BOUNDS IN $\mathbb{R}^2$

Lemma 4.7. *The variance $d^2(CH(P), l_{b_1})$ is bounded from below by $\frac{10}{243}Ab^2$.*

Proof. We split the integral $\int_0^b (y-b_1)^2 g(y)dy$ at b_1 (recall that b_1 is the y-coordinate of the center of gravity of $CH(P)$), and prove lower bounds on both parts in the following way: For the left part consider the linear function $f_1(y) = \frac{h_1}{b_1}y$ such that $\int_0^{b_1} f_1(y)dy = \int_0^{b_1} g(y)dy = A_1$ (see Figure 4.9 (a) for an illustration). From $\int_0^{b_1} f_1(y)dy = A_1$, it follows that $f_1(y) = \frac{2A_1 y}{b_1^2}$. Since $g(y)$ is convex, $g(y)$ and $f_1(y)$ intersect only once, at a point $b' \in (0, b_1)$.

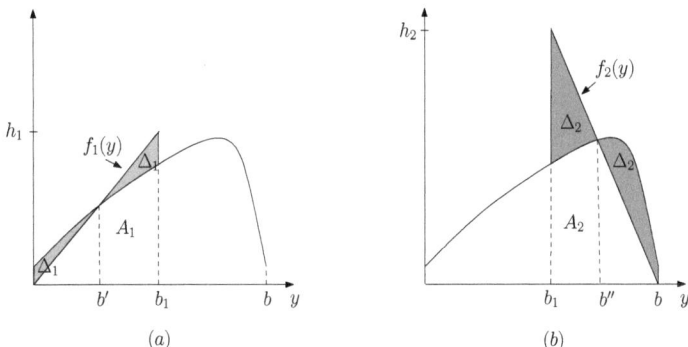

Figure 4.9: Construction of the lower bound on $d^2(CH(P), l_{b_1})$.

By Theorem 4.4, we have

$$\int_0^{b_1}(y-b_1)^2 g(y)dy \geq \int_0^{b_1}(y-b_1)^2 f_1(y)dy = \int_0^{b_1}(y-b)^2 \frac{2A_1}{b_1^2}dy = \frac{A_1 b_1^2}{6}. \tag{4.8}$$

Analogously, for the right part consider the linear function $f_2(y) = \frac{h_2}{b_1 - b}(y - b) = \frac{h_2}{-b_2}(y - b)$ such that $\int_{b_1}^b f_2(y)dy = \int_{b_1}^b g(y)dy = A_2$ (see Figure 4.9 (b) for an illustration). From $\int_{b_1}^b f_2(y)dy = A_2$, it follows that $f_2(y) = \frac{2A_2}{b_2^2}(y - b)$. Since $g(y)$ is convex, $g(y)$ and $f_2(y)$ intersect only once, at a point $b'' \in (b_1, b)$. By Theorem 4.4, we have that

$$\begin{aligned}\int_{b_1}^b (y-b_1)^2 g(y) dy &\geq \int_{b_1}^b (y-b_1)^2 f_2(y) dy \\ &= \int_{b_1}^b (y-b_1)^2 \tfrac{2A_2}{(b-b_1)^2}(y-b_1) dy \\ &= \tfrac{A_2 b_2^2}{6}.\end{aligned} \quad (4.9)$$

From (4.8) and (4.9) we obtain that

$$d^2(CH(P), l_{b_1}) = \int_0^{b_1}(y-b_1)^2 g(y) dy + \int_{b_1}^b (y-b_1)^2 g(y) dy \geq \tfrac{A_1 b_1^2}{6} + \tfrac{A_2 b_2^2}{6}.$$

From the Grünbaum-Hammer-Mityagin theorem, we know that $A_1, A_2 \in [\tfrac{4}{9}A, \tfrac{5}{9}A]$. Also, we know that $b_1, b_2 \in [\tfrac{1}{3}b, \tfrac{2}{3}b]$. It is not hard to show that, under these constrains, the expression $\tfrac{A_1 b_1^2}{6} + \tfrac{A_2 b_2^2}{6}$ achieves its minimum of $\tfrac{10}{243}Ab^2$ for $A_1 = \tfrac{4}{9}A, b_1 = \tfrac{5}{9}b$ or $A_1 = \tfrac{5}{9}A, b_1 = \tfrac{4}{9}b$. □

Lemma 4.8. *The variance $d^2(CH(P), l_{b_1})$ is bounded from above by $\tfrac{29}{243}Ab^2$.*

Proof. Without loss of generality, we can assume that $g(y)$ has its maximum in $[b_1, b]$. We split the integral $\int_0^b (y-b_1)^2 g(y) dy$ at b_1, and prove upper bounds for both parts in the following way. For the left part consider a linear function $f_3(y) = h_3$ such that $\int_0^{b_1} f_3(y) dy = \int_0^{b_1} g(y) dy = A_1$ (see Figure 4.10 (a) for an illustration). This implies

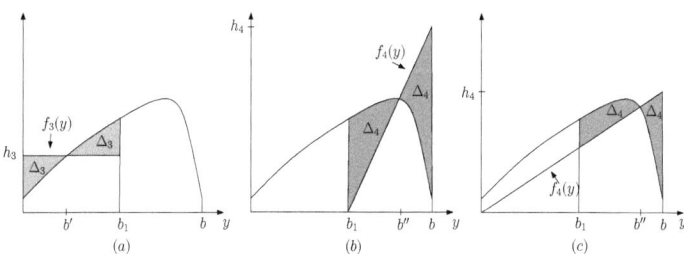

Figure 4.10: Construction of the upper bound for $d^2(CH(P), l_{b_1})$.

4.1. UPPER BOUNDS IN $\mathbb{R}^2$

that $f_3(y) = \frac{A_1}{b_1}$, and since $g(y)$ is convex, $g(y)$ and $f_3(y)$ intersect only once, at a point $b' \in (b_1, b)$. By Theorem 4.4, we have

$$\int_0^{b_1}(y-b_1)^2 g(y)dy \leq \int_0^{b_1}(y-b_1)^2 f_3(y)dy = \int_0^{b_1}(y-b_1)^2 \frac{A_1}{b_1} dy = \frac{A_1 b_1^2}{3}. \tag{4.10}$$

Now, we are looking for an appropriate function $f_4(y)$ to derive an upper bound of the second part of the integral $\int_0^b (y-b_1)^2 g(y)dy$. Note that both functions $f_3(y)$ and $f_4(y)$, in general can not be of the type $f(y) = const$, because it can happen that $f_4(y)$ intersects $g(y)$ twice, and we can not apply Theorem 4.4. Thus, for the left part we consider a linear function $f_4(y) = \frac{h_2}{b} y$ such that $\int_{b_1}^b f_4(y)dy = \int_{b_1}^b g(y)dy = A_2$ (see Figure 4.10 (b) for an illustration). $\int_{b_1}^b f_4(y)dy = A_2$ implies that $f_4(y) = \frac{2A_2 b_1}{b_2(b_1+b)} y$, and since $g(y)$ is convex, $g(y)$ and $f_4(y)$ intersect only once, at a point $b'' \in (b_1, b)$. By Theorem 4.4, we have

$$\int_{b_1}^b (y-b_1)^2 g(y)dy \geq \int_{b_1}^b (y-b_1)^2 f_4(y)dy = \int_{b_1}^b (y-b_1)^2 \frac{2A_2 b_1}{b_2(b_1+b)} y\, dy$$
$$= \frac{A_2 b_2^2}{b_1+b}\left(\frac{b_1}{4} + \frac{b_2}{4}\right). \tag{4.11}$$

From (4.10) and (4.11) we obtain

$$d^2(\mathcal{P}, l_{b_1}) = \int_0^{b_1}(y-b)^2 g(y)dy + \int_{b_1}^b (y-b)^2 g(y)dy$$
$$\leq \frac{A_1 b_1^2}{3} + \frac{A_2 b_2^2}{b_1+b}$$

From the Grünbaum-Hammer-Mityagin theorem, we know that $A_1, A_2 \in [\frac{4}{9}A, \frac{5}{9}A]$. Also, we know that $b_1, b_2 \in [\frac{1}{3}b, \frac{2}{3}b]$. It is not hard to show that, under these constrains, the expression $\frac{A_1 b_1^2}{3} + \frac{A_2 b_2^2}{b_1+b}\left(\frac{b_1+b_2}{4}\right)$ achieves its maximum of $\frac{29}{243}Ab^2$ for $A_1 = \frac{4}{9}A, b_1 = \frac{1}{3}b$. $\square$

We remark that in Lemma 4.8 we can use the function $f_4(y) =$

$\frac{h_4}{b_2}(y - b_1)$ instead of $f_4(y) = \frac{h_2}{b}y$ (see Figure 4.11 for an illustration), but that will give us bigger upper bound for $d^2(CH(P), l_{b_1})$, namely $\frac{34}{243}Ab^2$.

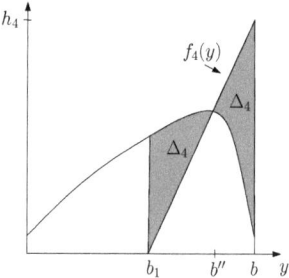

Figure 4.11: Construction of the upper bound for $d^2(CH(P), l_{b_1})$.

Now, we are ready to derive an alternative parameterized upper bound on $\kappa_{2,2}(P)$ which is better than the bound from Lemma 4.1 for big values of η.

Lemma 4.9. $\kappa_{2,2}(P) \leq \sqrt{2.9\left(1 + \frac{1}{\eta^2}\right)}$ for any point set P with aspect ratio $\eta(P) = \eta$.

Proof. Applying Lemma 4.7 and Lemma 4.8 in (4.6) we obtain
$$\frac{10}{243}Ab_{pca}^2 \leq d^2(\mathcal{P}, l_{pca}) \leq d^2(\mathcal{P}, l_{opt}) \leq \frac{29}{243}Ab_{opt}^2. \qquad (4.12)$$
From (4.12) it follows that $\beta = \frac{b_{pca}}{b_{opt}} \leq \sqrt{2.9}$. We have for a_{pca} the upper bound $diam(P) \leq \sqrt{a_{opt}^2 + b_{opt}^2} = a_{opt}\sqrt{1 + \frac{1}{\eta^2}}$. From this, it follows that $\alpha \leq \sqrt{1 + \frac{1}{\eta^2}}$. Putting this together, we obtain $\alpha\beta \leq \sqrt{2.9\left(1 + \frac{1}{\eta^2}\right)}$. □

Theorem 4.6. *The PCA bounding box of a point set P in $\mathbb{R}^2$ computed over $CH(P)$ has a guaranteed approximation factor $\kappa_{2,2} \leq 2.104$.*

4.2. AN UPPER BOUND IN $\mathbb{R}^3$

Proof. The theorem follows from the combination of the two parameterized bounds from Lemma 4.1 and Lemma 4.9:

$$\kappa_{2,2} \leq \sup_{\eta \geq 1} \left\{ \min\left(\eta + \frac{1}{\eta}, \sqrt{2.9\left(1 + \frac{1}{\eta^2}\right)}\right) \right\}.$$

It is easy to check that the supremum $s \approx 2.1038$ is obtained for $\eta \approx 1.3784$. □

4.1.4 An improved upper bound on $\kappa_{2,2}$

An improvement of the upper bound on $\kappa_{2,2}$ is obtained by applying the better upper bound on $\kappa_{2,2}$ for big η from Lemma 4.6 instead of applying the bound from Lemma 4.1.

Theorem 4.7. *The PCA bounding box of a point set P in $\mathbb{R}^2$ computed over $CH(P)$ has a guaranteed approximation factor $\kappa_{2,2} \leq 2.0695$.*

Proof. The theorem follows from the combination of the two parameterized bounds from Lemma 4.6 and Lemma 4.9:

$$\kappa_{2,2} \leq \sup_{\eta \geq 1} \left\{ \min\left(\frac{(\eta+1)^2}{2\eta}, \sqrt{2.9\left(1 + \frac{1}{\eta^2}\right)}\right) \right\}.$$

The supremum of the above expression occurs at 2.0694044 for $\eta \approx 1.4483691$. □

4.2 An upper bound in $\mathbb{R}^3$

4.2.1 An upper bound on $\kappa_{3,3}$

Some of the techniques used here are similar to those used in Subsection 4.1.3 where we derive an upper bound on $\kappa_{2,2}$. One essential difference is that for the upper bound on $\kappa_{3,3}$, we additionally need a bound for the ratio of the middle sides of $BB_{pca(3,3)}(P)$ and $BB_{opt}(P)$, which we derive from the relation in Lemma 4.13.

Given a point set $P \subseteq \mathbb{R}^3$ and an arbitrary bounding box $BB(P)$, we will denote the three side lengths of $BB(P)$ by a,b and c, where $a \geq b \geq c$. We are interested in the side lengths $a_{opt} \geq b_{opt} \geq c_{opt}$ and $a_{pca} \geq b_{pca} \geq c_{pca}$ of $BB_{opt}(P)$ and $BB_{pca(3,3)}(P)$. The parameters $\alpha = \alpha(P) = a_{pca}/a_{opt}$, $\beta = \beta(P) = b_{pca}/b_{opt}$ and $\gamma = \gamma(P) = c_{pca}/c_{opt}$ denote the ratios between the corresponding side lengths. Hence, we have $\kappa_{3,3}(P) = \alpha \cdot \beta \cdot \gamma$.

Since the side lengths of any bounding box are bounded by the diameter of P, we can observe that in general $c_{pca} \leq b_{pca} \leq a_{pca} \leq diam(P) \leq \sqrt{3} a_{opt}$, and in the special case when the optimal bounding box is a cube $\kappa_{3,3}(P) \leq 3\sqrt{3}$. This observation can be generalized, introducing two additional parameters $\eta(P) = a_{opt}/b_{opt}$ and $\theta(P) = a_{opt}/c_{opt}$.

Lemma 4.10. $\kappa_{3,3}(P) \leq \eta\,\theta \left(1 + \frac{1}{\eta^2} + \frac{1}{\theta^2}\right)^{\frac{3}{2}}$ for any point set P with aspect ratios $\eta(P) = \eta$ and $\theta(P) = \theta$.

Proof. We have for the side lengths a_{pca}, b_{pca} and c_{pca} the upper bound $diam(P) \leq \sqrt{a_{opt}^2 + b_{opt}^2 + c_{opt}^2} = a_{opt}\sqrt{1 + \frac{1}{\eta^2} + \frac{1}{\theta^2}}$. Thus, $\alpha\,\beta\,\gamma \leq \frac{a_{pca}\,b_{pca}\,c_{pca}}{a_{opt}\,b_{opt}\,c_{opt}} \leq \frac{a_{opt}^3 \left(1 + \frac{1}{\eta^2}\right)^{\frac{3}{2}}}{a_{opt}\,b_{opt}\,c_{opt}}$. Replacing a_{opt} in the nominator once by $\eta\,b_{opt}$ and once by $\theta\,c_{opt}$ we obtain $\kappa_{3,3}(P) \leq \eta\,\theta \left(1 + \frac{1}{\eta^2} + \frac{1}{\theta^2}\right)^{\frac{3}{2}}$. □

Unfortunately, this parameterized upper bound tends to infinity for $\eta \to \infty$ or $\theta \to \infty$. Therefore, we are going to derive another upper bound that is better for large values of η and θ. We derive such a bound by finding constants that bound β and γ from above. In this process we will make essential use of the properties of $BB_{pca(3,3)}(P)$. We denote by $d^2(CH(P), H)$ the integral of

4.2. AN UPPER BOUND IN $\mathbb{R}^3$

the squared distances of the points on $CH(P)$ to a plane H, i.e., $d^2(CH(P), H) = \int_{s \in CH(P)} d^2(s, H) ds$. Let H_{pca} be the plane going through the center of gravity, parallel to the side $a_{pca} \times b_{pca}$ of $BB_{pca(3,3)}(P)$, and H_{opt} be the bisector of $BB_{opt(P)}$ parallel to the side $a_{opt} \times b_{opt}$. By Proposition 2.5, H_{pca} is the best fitting plane of P and therefore,

$$d^2(CH(P), H_{pca}) \leq d^2(CH(P), H_{opt}). \tag{4.13}$$

We obtain an estimation for γ by determining a lower bound on $d^2(CH(P), H_{pca})$ that depends on c_{pca}, and an upper bound on $d^2(CH(P), H_{opt})$ that depends on c_{opt}. Having an arbitrary bounding box of $CH(P)$ (with side lengths a, b, and c, $a \geq b \geq c$), we denote by H_{ab} the plane going through the center of gravity, parallel to the side $a \times b$. The volume of $CH(P)$ can be expressed as

$$V = V(CH(P)) = \int_0^c \int_0^b \int_0^a \chi_{CH(P)}(x, y, z) dx dy dz = \int_0^c g(z) dz,$$

where $\chi_{CH(P)}(x, y, z)$ is the *characteristic function* of $CH(P)$ defined as

$$\chi_{CH(P)}(x, y, z) = \begin{cases} 1 & (x, y, z) \in CH(P) \\ 0 & (x, y, z) \notin CH(P), \end{cases}$$

and $g(z) = \int_0^b \int_0^a \chi_{CH(P)}(x, y, z) dx dy$ is the area of the intersection of $CH(P)$ with the horizontal plane at height z. As before we call $g(z)$ the density function of $CH(P)$. Let c_1 denote the z-coordinate of the center of gravity of $CH(P)$. The line l_{c_1} ($y = c_1$) divides the volume of $CH(P)$ into V_1 and V_2 (see Figure 4.12 (a) for an illustration).

Note that $g(z)$ is continuous, but in general not convex in the interval $[0, b]$. Therefore, we cannot use linear functions to derive a lower and an upper bound on the function $d^2(CH(P), H_{ab})$, as we did in Subsection 4.1.3, because a linear function can intersect $g(z)$ more

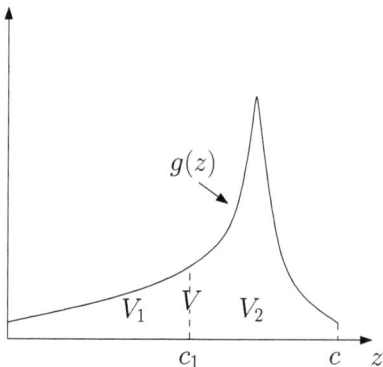

Figure 4.12: Density function $g(y)$ of the convex hull of a point set in $\mathbb{R}^3$.

than once, and we cannot apply Theorem 4.4. We will show that instead of linear functions, quadratic functions can be used.

Proposition 4.1. *Let $g(z)$ be the density function of $CH(P)$ defined as above, and let $f(z) = kz^2$ be the parabola such that $\int_0^{c_1} f(z)dz = \int_0^{c_1} g(z)dz$. Then, $\exists c_0 \in [0, c_1]$ such that $f(z) \leq g(x)$ for all $z \leq c_0$ and $f(z) \geq g(z)$ for all $z \geq c_0$.*

Proof. We give a constructive proof. Let $c_0 := \inf \{ d \mid \forall z \in [d, c_1]\, g(z) \leq f(z)\}$. If $c_0 = 0$, then $f(z) = g(z)$, and the proposition holds. If $c_0 > 0$, then consider the polygon which is the intersection of $CH(P)$ with the plane $z = c_0$. We fix a point p_0 in $CH(P)$ with z-coordinate 0 and construct a pyramid Q by extending all rays from p_0 through the polygon up to the plane $z = c_1$ (see Figure 4.13 for an illustration). Since, $f(c_0) = g(c_0)$ the quadratic function $f(z)$ is the density function of Q. Therefore, since the part of Q below c_0 is completely included in $CH(P)$, we can conclude that $f(z) \leq g(z)$ for all $z \leq c_0$. On the other hand, $f(z) \geq g(x)$ for all $z \geq c_0$ by the definition of c_0. □

Now, we present a lower and an upper bound on the variance

4.2. AN UPPER BOUND IN $\mathbb{R}^3$

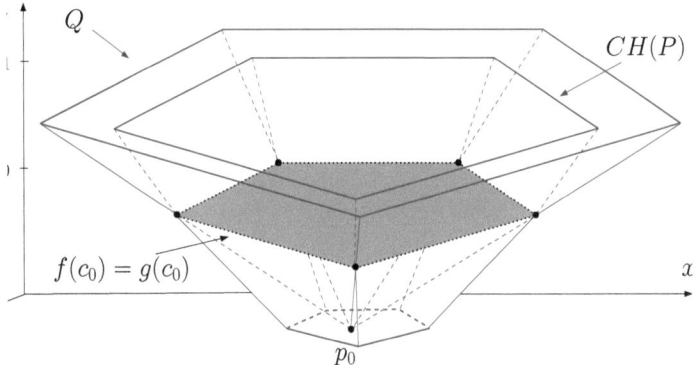

Figure 4.13: Construction of the intersection of $f(z)$ and $g(z)$.

$d^2(CH(P), H_{ab})$, from which we can derive a bound on $\gamma = \frac{c_{pca}}{c_{opt}}$.

Lemma 4.11. *The variance $d^2(CH(P), H_{ab})$ is bounded from below by $\frac{7}{256}Vc^2$.*

Proof. We split the integral $\int_0^c (z-c_1)^2 g(z) dz$ at c_1, and prove upper bounds on both parts in the following way: For the left part consider the parabola $f_1(z) = \frac{h_1}{c_1^2} z^2$ such that $\int_0^{c_1} f_1(z) dz = \int_0^{c_1} g(z) dz = V_1$ (see Figure 4.14 (a) for an illustration). From $\int_0^{c_1} f_1(z) dz = V_1$ we have that $f_1(z) = \frac{3V_1}{c_1^3} z^2$. Since $f_1(z)$ and $g(z)$ define the same volume on the interval $[0, c_1]$, they must intersect, and by Proposition 4.1 we know that if $f_1(z) \neq g(z)$, then they can intersect only once, at a point $c' \in (0, c_1)$. Under these conditions, we can apply Theorem 4.4, and obtain

$$\int_0^{c_1}(z-c_1)^2 g(z) dz \geq \int_0^{c_1}(z-c_1)^2 f_1(z) dz = \int_0^{c_1}(z-c_1)^2 \frac{3V_1}{c_1^3} z^2 dz = \frac{V_1 c_1^2}{10}. \quad (4.14)$$

Analogously, for the right part consider the parabola $f_2(z) = \frac{h_2}{(c_1-c)^2}(z-c)^2 = \frac{h_2}{c_2^2}(z-c)^2$ such that $\int_{c_1}^c f_2(y) dy = \int_{c_1}^c g(z) dz = V_2$ (see Figure 4.14 (b) for an illustration). From $\int_{c_1}^{c} f_2(y)dy = V_2$ we have that $f_1(z) = \frac{3V_2}{c_2^3}(z-c)^2$. By similar arguments as above in the case of

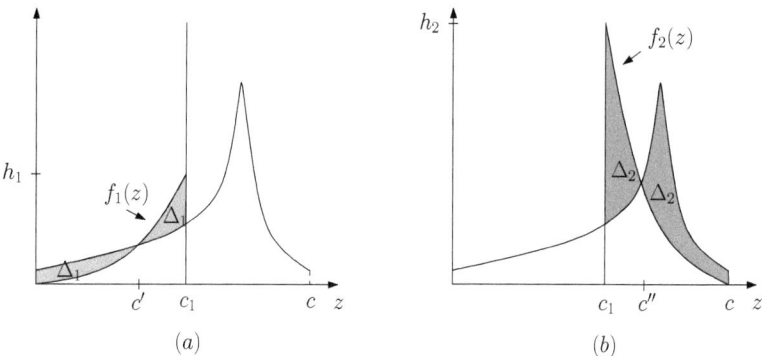

Figure 4.14: Construction of the lower bound on $d^2(CH(P), H_{ab})$

$f_1(z)$, we can show that $g(z)$ and $f_2(z)$ intersect only once, at a point $c'' \in (c_1, c)$. Applying Theorem 4.4 we have that

$$\begin{aligned}\int_{c_1}^{c}(z-c_1)^2 g(z)dz &\geq \int_{c_1}^{c}(z-c_1)^2 f_2(z)dz \\ &= \int_{c_1}^{c}(z-c_1)^2 \frac{3V_2}{c_2^3}(z-c)^2 dz \quad (4.15)\\ &= \frac{V_2 c_2^2}{10}.\end{aligned}$$

From (4.14) and (4.15) we obtain that

$$\begin{aligned}d^2(CH(P), H_{ab}) &= \int_0^{c_1}(z-c_1)^2 g(z)dz + \int_{c_1}^{c}(z-c_1)^2 g(z)dz \\ &\geq \frac{V_1 c_1^2}{10} + \frac{V_2 c_2^2}{10}.\end{aligned}$$

From the Grünbaum-Hammer-Mityagin theorem, we know that $V_1, V_2 \in [\frac{27}{64}V, \frac{37}{64}V]$. Also, we know that $c_1, c_2 \in [\frac{1}{4}c, \frac{3}{4}c]$. It is not hard to show that, under these constrains, the expression $\frac{V_1 c_1^2}{10} + \frac{V_2 c_2^2}{10}$ achieves its minimum of $\frac{7}{256}Vc^2$ for $V_1 = \frac{27}{64}V, c_1 = \frac{3}{4}c$ or $V_1 = \frac{37}{64}V, c_1 = \frac{1}{4}c$. □

4.2. AN UPPER BOUND IN $\mathbb{R}^3$

Lemma 4.12. *The variance $d^2(CH(P), H_{ab})$ is bounded from above by $\frac{12729}{71680} V c^2$.*

Proof. Without loss of generality, we can assume that $g(z)$ has its maximum in $[c_1, c]$. We split the integral $\int_0^c (z - c_1)^2 g(z) dz$ at c_1, and prove upper bounds for both parts in the following way: For the left part consider the linear function $f_3(z) = h_3$ such that $\int_0^{c_1} f_3(z) dz = \int_0^{c_1} g(z) dz = V_1$ (see Figure 4.15 (a) for an illustration).

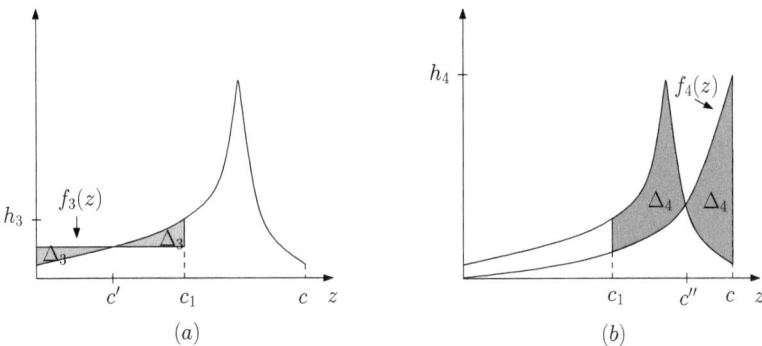

Figure 4.15: Construction of the upper bound on $d^2(CH(P), H_{ab})$

From $\int_0^{c_1} f_3(z) dz = V_1$ we have that $f_3(z) = \frac{V_1}{c_1}$. Since $f_3(z)$ is constant, it intersects $g(z)$ only once, at a point $c' \in (c_1, c)$. By Theorem 4.4, we have that

$$\int_0^{c_1} (z - c_1)^2 g(z) dz \leq \int_0^{c_1} (z - c_1)^2 f_3(z) dz = \int_0^{c_1} (z - c_1)^2 \frac{V_1}{c_1} dz = \frac{V_1 c_1^2}{3}. \tag{4.16}$$

Now, we are looking for an appropriate function $f_4(z)$ to derive an upper bound on the second part of the integral $\int_0^z (z - c_1)^2 g(z) dz$. Note that both functions $f_3(z)$ and $f_4(z)$, in general can not be of the type $f(y) = const$, which give us the best upper bound, because it can happen that $f_4(z)$ intersects $g(z)$ twice, and we can not apply Theorem 4.4. Thus, for the right part we consider the parabola $f_4(z) = \frac{h_4}{c^2}z^2$ such that $\int_{c_1}^c f_4(z)dz = \int_{c_1}^c g(z)dz = V_2$ (see Figure 4.15 (b) for an illustration). Since $f_4(z)$ and $g(z)$ define the same volume on the interval $[c_1, c]$, they must intersect, and by Proposition 4.1 we know that if $f_4(z) \neq g(z)$, they can intersect only once, at a point $c' \in (c_1, c)$. Under these conditions, we can apply Theorem 4.4, and since $f_4(z) = \frac{3V_2}{c^3-c_1^3}z^2$, we obtain

$$\int_{c_1}^c (z-c_1)^2 g(z)dz \geq \int_{c_1}^c (z-c_1)^2 f_4(z)dz = \int_{c_1}^c (z-c_1)^2 \frac{3V_2}{c^3-c_1^3}z^2 dz$$

$$= \frac{3V_2 c_2^2}{c^2+c c_1+c_1^2}\left(\frac{c_2^2}{5} + \frac{c_2 c_1}{2} + \frac{c_1^3}{3}\right). \quad (4.17)$$

From (4.16) and (4.17) we can conclude that

$$d^2(\mathcal{P}, H_{ab}) = \int_0^{c_1}(z-c)^2 g(z)dz + \int_{c_1}^c (z-c)^2 g(z)dz$$

$$\leq \frac{V_1 c_1^2}{3} + \frac{3V_2 c_2^2}{c^2+c c_1+c_1^2}\left(\frac{c_2^2}{5} + \frac{c_2 c_1}{2} + \frac{c_1^3}{3}\right).$$

From the Grünbaum-Hammer-Mityagin theorem, we know that $V_1, V_2 \in [\frac{27}{64}V, \frac{37}{64}V]$. Also, we know that $c_1, c_2 \in [\frac{1}{4}c, \frac{3}{4}c]$. It is not hard to show that, under these constrains, the expression $\frac{V_1 c_1^2}{3} + \frac{3V_2 c_2^2}{c^2+c c_1+c_1^2}\left(\frac{c_2^2}{5} + \frac{c_2 c_1}{2} + \frac{c_1^3}{3}\right)$ achieves its maximum of $\frac{12729}{71680}Vc^2$ for $V_1 = \frac{27}{64}V$, $c_1 = \frac{1}{4}c$. □

We remark that in Lemma 4.12 we can use the function $f_4(z) = \frac{h_4}{(c-c_1)^2}(z-c_1)^2$ instead of $f_4(z) = \frac{h_4}{c^2}z^2$ (see Figure 4.16 for an illustration), but that will give us bigger upper bound for $d^2(CH(P), H_{ab})$, namely $\frac{3132}{15360}Vc^2$.

As a consequence of Lemma 4.11 and Lemma 4.12, we have the following upper bound on γ.

Proposition 4.2. $\gamma < 2.5484$.

4.2. AN UPPER BOUND IN $\mathbb{R}^3$

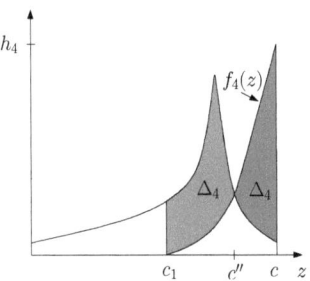

Figure 4.16: Construction of an alternative upper bound on $d^2(CH(P), H_{ab})$

Proof. By Lemma 4.11, we have

$$\frac{7}{256} V c_{pca}^2 \leq d^2(CH(P), H_{pca}). \tag{4.18}$$

On the other hand, by Lemma 4.12, it follows that

$$d^2(CH(P), H_{opt}) \leq \frac{12729}{71680} V c_{opt}^2, \tag{4.19}$$

From (4.18), (4.19) and (4.13), we obtain

$$\gamma = \frac{c_{pca}}{c_{opt}} \leq \sqrt{\frac{12729}{1960}} < 2.5484.$$

□

We are now ready to present a new parameterized bound on $\kappa_{3,3}(P)$, which is good for large values of η and θ. The additional crucial relation we exploit in its derivation is the fact given in the following lemma.

Lemma 4.13. *Let $(x_1, x_2, \ldots, x_d)$ and $(y_1, y_2, \ldots, y_d)$ be two sets of orthogonal base vectors in $\mathbb{R}^d$. For any point set $P \in \mathbb{R}^d$ it holds that*

$$\sum_{i=1}^{d} var(P, x_i) = \sum_{i=1}^{d} var(P, y_i).$$

Proof. We have that

$$\sum_{i=1}^{d} \mathrm{var}(P, x_i) = \sum_{i=1}^{d} \frac{1}{n} \sum_{p \in P} d^2(p, H_{x_i}),$$

where H_{x_i} is a hyperplane orthogonal to the vector x_i, passing through the origin of the coordinate system, $d^2(p, H_{x_i})$ denotes the Euclidean distance of p to H_{x_i}, and $n = |P|$. Since $\sum_{i=1}^{d} d^2(p, H_{x_i})$ is the squared distance of p to the origin of the coordinate system, it can be expressed as the sum of squared distances to the $(d-1)$-dimensional hyperplanes spanned by any set of orthogonal base vectors. Therefore,

$$\sum_{i=1}^{d} d^2(p, H_{x_i}) = \sum_{i=1}^{d} d^2(p, H_{y_i}), \quad \text{and}$$

$$\begin{aligned}
\sum_{i=1}^{d} \mathrm{var}(P, x_i) &= \tfrac{1}{n} \sum_{p \in P} \sum_{i=1}^{d} d^2(p, H_{x_i}) \\
&= \tfrac{1}{n} \sum_{p \in P} \sum_{i=1}^{d} d^2(p, H_{y_i}) \\
&= \sum_{i=1}^{d} \mathrm{var}(P, y_i).
\end{aligned}$$

When P is a continuous point set,

$$\mathrm{var}(P, x_i) = \frac{1}{\mathrm{Vol}(P)} \int_{p \in P} d^2(p, H_{x_i}) ds$$

and the claim can be shown as in the discrete case. □

Lemma 4.14. $\kappa_{3,3}(P) \leq 6.43 \sqrt{1 + \frac{1}{\eta^2} + \frac{1}{\theta^2}}$ *for any point set P with aspect ratios $\eta(P) = \eta$ and $\theta(P) = \theta$.*

Proof. Let $x_{pca}, y_{pca}, z_{pca}$ be a set of basis vectors that determine the direction of $BB_{pca(3,3)}(P)$, and let $x_{opt}, y_{opt}, z_{opt}$ be a set of basis vectors that determine the direction of $BB_{opt}(CH(P))$. By Lemma 4.13, we

4.2. AN UPPER BOUND IN $\mathbb{R}^3$

have that

$$\text{var}(CH(P), x_{pca}) + \text{var}(CH(P), y_{pca}) + \text{var}(CH(P), z_{pca}) =$$
$$\text{var}(CH(P), x_{opt}) + \text{var}(CH(P), y_{opt}) + \text{var}(CH(P), z_{opt}). \quad (4.20)$$

By Proposition 2.2, the variance of $CH(P)$ in the direction x_{pca} is the biggest possible, and therefore,

$$\text{var}(CH(P), x_{pca}) \geq \text{var}(CH(P), x_{opt}). \quad (4.21)$$

Combining (4.20) and (4.21) we obtain

$$\text{var}(CH(P), y_{pca}) + \text{var}(CH(P), z_{pca}) \leq$$
$$\text{var}(CH(P), y_{opt}) + \text{var}(CH(P), z_{opt}). \quad (4.22)$$

We denote by $H_{a_p b_p}$ the plane orthogonal to z_{pca}, going through the center of gravity, and parallel with the side $a_{pca}b_{pca}$ of $BB_{pca(3,3)}(P)$. Similarly, we define $H_{a_p c_p}$, $H_{a_o b_o}$ and $H_{a_o c_o}$. We can rewrite (4.22) as

$$d^2(CH(P), H_{a_p b_p}) + d^2(CH(P), H_{a_p c_p}) \leq$$
$$d^2(CH(P), H_{a_o b_o}) + d^2(CH(P), H_{a_o c_o}). \quad (4.23)$$

By Lemma 4.11, the lower bound on $d^2(CH(P), H_{a_p b_p})$ is $\frac{7}{256}Vc_{pca}^2$, and the lower bound on $d^2(CH(P), H_{a_p c_p})$ is $\frac{7}{256}Vb_{pca}^2$. By Lemma 4.12, the upper bound on $d^2(CH(P), H_{a_o b_o})$ is $\frac{12729}{71680}Vc_{opt}^2$, and the lower bound on $d^2(CH(P), H_{a_o c_o})$ is $\frac{12729}{71680}Vb_{opt}^2$. Plugging these bounds into (4.23) we obtain

$$\frac{7}{256}Vc_{pca}^2 + \frac{7}{256}Vb_{pca}^2 \leq \frac{12729}{71680}Vc_{opt}^2 + \frac{12729}{71680}Vb_{opt}^2. \quad (4.24)$$

Applying $\gamma = \frac{c_{pca}}{c_{opt}}$ in (4.24), we obtain

$$\frac{7}{256}b_{pca}^2 \leq \left(\frac{12729}{71680} - \frac{7}{256}\gamma\right)c_{opt}^2 + \frac{12729}{71680}b_{opt}^2. \quad (4.25)$$

By Proposition 4.2, it follows that $\frac{12729}{71680} - \frac{7}{256}\gamma \geq 0$, and since $b_{opt} \geq c_{opt}$, we get from (4.25) that

$$\beta = \frac{b_{pca}}{b_{opt}} \leq \sqrt{12.99 - \gamma^2}. \tag{4.26}$$

The expression $\sqrt{12.99 - \gamma^2}\,\gamma\ (\geq \beta\gamma)$ has its maximum of 6.495 for $\gamma \approx 2.5484$. This together with the bound $\alpha \leq \sqrt{1 + \frac{1}{\eta^2} + \frac{1}{\theta^2}}$ gives

$$\kappa_{3,3}(P) = \alpha\,\beta\,\gamma \leq 6.495\sqrt{1 + \frac{1}{\eta^2} + \frac{1}{\theta^2}}.$$

$\square$

Lemma 4.10 gives us a bound on $\kappa_{3,3}(P)$ which is good for small values of η and θ. In contrary, the bound from Lemma 4.14 behaves worse for small values of η and θ, but better for big values of η and θ. Therefore, we combine both of them to obtain the final upper bound.

Theorem 4.8. *The PCA bounding box of a point set P in $\mathbb{R}^3$ computed over $CH(P)$ has a guaranteed approximation factor $\kappa_{3,3} < 7.81$.*

Proof. The theorem follows from the combination of the two parameterized bounds from Lemma 4.10 and Lemma 4.14:

$$\kappa_{3,3} \leq \sup\nolimits_{\eta \geq 1,\,\theta \geq 1} \left\{ \min\left(\eta\,\theta\left(1 + \frac{1}{\eta^2} + \frac{1}{\theta^2}\right)^{\frac{3}{2}}, 6.495\sqrt{1 + \frac{1}{\eta^2} + \frac{1}{\theta^2}}\right)\right\}.$$

By numerical verification we obtained that the supremum occurs at ≈ 7.8073 for $\eta = 2.12$ and $\theta \approx 2.1203$. $\square$

4.3 Open Problems

Improving the upper bound on $\kappa_{3,3}$, $\kappa_{2,2}$ and $\kappa_{2,1}$, as well as obtaining an upper bound on $\kappa_{3,2}$ is of interest. The approaches for obtaining the upper bounds exploit in this chapter requires an estimation of

4.3. OPEN PROBLEMS

the length ratio between each corresponding sides of the minimum-volume bounding box and the PCA bounding box. However, even in $\mathbb{R}^4$, we do not know how to obtain the estimations of the length ratio between all corresponding sides. We believe that obtaining upper bounds on the approximation factor on the quality of PCA bounding boxes in arbitrary dimension requires different approaches than those presented here.

An interesting open problem on its own is to find the maximum-volume bounding box in $\mathbb{R}^3$ whose sides touch a bounding box with predefined side lengths. We expect that, similarly as in $\mathbb{R}^2$, this will lead to a better upper bound for big η and θ than the bound obtained in Lemma 4.10.

Chapter 5
Closed-form Solutions for Continuous PCA

In this chapter, we consider the continuous version of PCA, and give the closed form solutions for the case when the point set is a convex polygon or boundary of the convex polygon in $\mathbb{R}^2$, or polyhedron or a polyhedral surface in $\mathbb{R}^3$. To the best of our knowledge, it is the first time that the continuous PCA over the volume of the 3D body has been considered. Closed-form solutions of variants of the continuous PCA of a polyhedral surface can be found in [14, 50].

5.1 Evaluation of the Expressions for Continuous PCA

Although the continuous PCA approach is based on integrals, it is possible to reduce the formulas to ordinary sums if the point set X in $\mathbb{R}^2$ is a polygon or boundary of a polygon. Closed-form solutions are presented also if the point set X in $\mathbb{R}^3$ is a polyhedron or a polyhedral surface.

5.1.1 Continuous PCA in $\mathbb{R}^2$

Continuous PCA over a polygon

We assume that the polygon X is triangulated (if it is not, we can triangulate it in preprocessing), and the number of triangles is n. The i-th triangle, with vertices $\vec{x}_{1,i}, \vec{x}_{2,i}, \vec{x}_{3,i} = \vec{o}$, can be represented in a parametric form by $\vec{T}_i(s,t) = \vec{x}_{3,i} + s(\vec{x}_{1,i} - \vec{x}_{3,i}) + t(\vec{x}_{2,i} - \vec{x}_{3,i})$, for $0 \leq s, t \leq 1$, and $s + t \leq 1$.

The center of gravity of the i-th triangle is

$$\vec{c}_i = \frac{\int_0^1 \int_0^{1-s} \vec{T}_i(s,t)\, dt\, ds}{\int_0^1 \int_0^{1-s} dt\, ds} = \frac{\vec{x}_{1,i} + \vec{x}_{2,i} + \vec{x}_{3,i}}{3}.$$

The contribution of each triangle to the center of gravity of X is proportional to its area. The area of the i-th triangle is

$$a_i = \text{area}(T_i) = \frac{|(\vec{x}_{2,i} - \vec{x}_{1,i})| \times |(\vec{x}_{3,i} - \vec{x}_{1,i})|}{2},$$

where $\times$ denotes the vector product. We introduce a weight to each triangle that is proportional with its area, define as

$$w_i = \frac{a_i}{\sum_{i=1}^n a_i}.$$

Then, the center of gravity of X is

$$\vec{c} = \sum_{i=1}^n w_i \vec{c}_i.$$

The covariance matrix of the i-th triangle is

$$\begin{aligned} C_i &= \frac{\int_0^1 \int_0^{1-s} (\vec{T}_i(s,t) - \vec{c})(\vec{T}_i(s,t) - \vec{c})^T\, dt\, ds}{\int_0^1 \int_0^{1-s} dt\, ds} \\ &= \tfrac{1}{12}\Big(\sum_{j=1}^3 \sum_{k=1}^3 (\vec{x}_{j,i} - \vec{c})(\vec{x}_{k,i} - \vec{c})^T + \\ &\quad \sum_{j=1}^3 (\vec{x}_{j,i} - \vec{c})(\vec{x}_{j,i} - \vec{c})^T \Big). \end{aligned}$$

The element C_i^{ab} of C_i, where $a, b \in \{1, 2\}$ is

$$\begin{aligned} C_i^{ab} = \tfrac{1}{12}\Big(&\sum_{j=1}^3 \sum_{k=1}^3 (x_{j,i}^a - c^a)(x_{k,i}^b - c^b) + \\ &\sum_{j=1}^3 (x_{j,i}^a - c^a)(x_{j,i}^b - c^b) \Big), \end{aligned}$$

with $\vec{c} = (c^1, c^2)$. The covariance matrix of X is

$$C = \sum_{i=1}^n w_i C_i.$$

Continuous PCA over the boundary of a polygon

Let X be a polygon in $\mathbb{R}^2$. We assume that the boundary of X is comprised of n line segments. The i-th line segment, with vertices $\vec{x}_{1,i}, \vec{x}_{2,i}$, can be represented in a parametric form by

$$\vec{L}_i(s) = \vec{x}_{1,i} + s\,(\vec{x}_{2,i} - \vec{x}_{1,i}).$$

Since we assume that the mass density is constant, the center of gravity of the i-th line segment is

$$\vec{c}_i = \frac{\int_0^1 \vec{L}_i(s)\, ds}{\int_0^1 ds} = \frac{\vec{x}_{1,i} + \vec{x}_{2,i}}{2}.$$

The contribution of each line segment to the center of gravity of the boundary of a polygon is proportional with the length of the line segment. The length of the i-th line segment is

$$l_i = \text{length}(L_i) = ||\vec{x}_{2,i} - \vec{x}_{1,i}||.$$

We introduce a weight to each line segment that is proportional with its length, define as

$$w_i = \frac{l_i}{\sum_{i=1}^n l_i}.$$

Then, the center of gravity of the boundary of X is

$$\vec{c} = \sum_{i=1}^n w_i \vec{c}_i.$$

The covariance matrix of the i-th line segment is

$$\begin{aligned}
C_i &= \frac{\int_0^1 (\vec{L}_i(s) - \vec{c})(\vec{L}_i(s) - \vec{c})^T\, ds}{\int_0^1 ds} \\
&= \tfrac{1}{6}\Big(\sum_{j=1}^2 \sum_{k=1}^2 (\vec{x}_{j,i} - \vec{c})(\vec{x}_{k,i} - \vec{c})^T + \\
&\quad \sum_{j=1}^2 (\vec{x}_{j,i} - \vec{c})(\vec{x}_{j,i} - \vec{c})^T\Big).
\end{aligned}$$

The element C_i^{ab} of C_i, where $a, b \in \{1, 2\}$ is

$$C_i^{ab} = \tfrac{1}{6}\Big(\sum_{j=1}^{2}\sum_{k=1}^{2}(x_{j,i}^a - c^a)(x_{k,i}^b - c^b) + \sum_{j=1}^{2}(x_{j,i}^a - c^a)(x_{j,i}^b - c^b)\Big),$$

with $\vec{c} = (c^1, c^2)$. The covariance matrix of the boundary of X is

$$C = \sum_{i=1}^{n} w_i C_i.$$

5.1.2 Continuous PCA in $\mathbb{R}^3$

Continuous PCA over a (convex) polyhedron

Let X be a convex polytope in $\mathbb{R}^3$. We assume that the boundary of X is triangulated (if it is not, we can triangulate it in preprocessing). We choose an arbitrary point $\vec{o}$ in the interior of X, for example, we can choose that $\vec{o}$ is the center of gravity of the boundary of X. Each triangle from the boundary together with $\vec{o}$ forms a tetrahedron. Let the number of such formed tetrahedra be n. The i-th tetrahedron, with vertices $\vec{x}_{1,i}, \vec{x}_{2,i}, \vec{x}_{3,i}, \vec{x}_{4,i} = \vec{o}$, can be represented in a parametric form by $\vec{Q}_i(s,t,u) = \vec{x}_{4,i} + s(\vec{x}_{1,i} - \vec{x}_{4,i}) + t(\vec{x}_{2,i} - \vec{x}_{4,i}) + u(\vec{x}_{3,i} - \vec{x}_{4,i})$, for $0 \leq s, t, u \leq 1$, and $s + t + u \leq 1$.

The center of gravity of the i-th tetrahedron is

$$\vec{c}_i = \frac{\int_0^1 \int_0^{1-s} \int_0^{1-s-t} \rho(\vec{Q}_i(s,t))\vec{Q}_i(s,t)\,du\,dt\,ds}{\int_0^1 \int_0^{1-s} \int_0^{1-s-t} \rho(\vec{Q}_i(s,t))\,du\,dt\,ds},$$

where $\rho(\vec{Q}_i(s,t))$ is a mass density at a point $\vec{Q}_i(s,t)$. Since, we can assume $\rho(\vec{Q}_i(s,t)) = 1$, we have

$$\vec{c}_i = \frac{\int_0^1 \int_0^{1-s} \int_0^{1-s-t} \vec{Q}_i(s,t)\,du\,dt\,ds}{\int_0^1 \int_0^{1-s} \int_0^{1-s-t} du\,dt\,ds} = \frac{\vec{x}_{1,i} + \vec{x}_{2,i} + \vec{x}_{3,i} + \vec{x}_{4,i}}{4}.$$

The contribution of each tetrahedron to the center of gravity of X is proportional to its volume. If M_i is the 3×3 matrix whose k-th row

5.1. EVALUATION OF THE EXPRESSIONS FOR CONTINUOUS PCA

is $\vec{x}_{k,i} - \vec{x}_{4,i}$, for $k = 1\ldots 3$, then the volume of the i-th tetrahedron is

$$v_i = \text{volume}(Q_i) = \frac{|det(M_i)|}{3!}.$$

We introduce a weight to each tetrahedron that is proportional with its volume, define as

$$w_i = \frac{v_i}{\sum_{i=1}^n v_i}.$$

Then, the center of gravity of X is

$$\vec{c} = \sum_{i=1}^n w_i \vec{c}_i.$$

The covariance matrix of the i-th tetrahedron is

$$\begin{aligned} C_i &= \frac{\int_0^1 \int_0^{1-s} \int_0^{1-s-t} (\vec{Q}_i(s,t,u) - \vec{c})\,(\vec{Q}_i(s,t,u) - \vec{c})^T \, du\, dt\, ds}{\int_0^1 \int_0^{1-s} \int_0^{1-s-t} du\, dt\, ds} \\ &= \frac{1}{20}\Big(\sum_{j=1}^4 \sum_{k=1}^4 (\vec{x}_{j,i} - \vec{c})(\vec{x}_{k,i} - \vec{c})^T + \\ &\qquad \sum_{j=1}^4 (\vec{x}_{j,i} - \vec{c})(\vec{x}_{j,i} - \vec{c})^T \Big). \end{aligned}$$

The element C_i^{ab} of C_i, where $a, b \in \{1, 2, 3\}$ is

$$\begin{aligned} C_i^{ab} &= \frac{1}{20}\Big(\sum_{j=1}^4 \sum_{k=1}^4 (x_{j,i}^a - c^a)(x_{k,i}^b - c^b) + \\ &\qquad \sum_{j=1}^4 (x_{j,i}^a - c^a)(x_{j,i}^b - c^b) \Big), \end{aligned}$$

with $\vec{c} = (c^1, c^2, c^3)$. Finally, the covariance matrix of X is

$$C = \sum_{i=1}^n w_i C_i.$$

We would like to note that the above expressions hold also for a star-shaped object, where $\vec{o}$ is the kernel of the object, or for any non-convex tetrahedralized polyhedron.

Continuous PCA over a boundary of a polyhedron

Let X be a polyhedron in $\mathbb{R}^3$. We assume that the boundary of X is triangulated, containing n triangles. The i-th triangle, with vertices

$\vec{x}_{1,i}, \vec{x}_{2,i}, \vec{x}_{3,i}$, can be represented in a parametric form by $\vec{T}_i(s,t) = \vec{x}_{1,i} + s(\vec{x}_{2,i} - \vec{x}_{1,i}) + t(\vec{x}_{3,i} - \vec{x}_{1,i})$, for $0 \leq s, t \leq 1$, and $s + t \leq 1$.

The center of gravity of the i-th triangle is

$$\vec{c}_i = \frac{\int_0^1 \int_0^{1-s} \vec{T}_i(s,t)\, dt\, ds}{\int_0^1 \int_0^{1-s} dt\, ds} = \frac{\vec{x}_{1,i} + \vec{x}_{2,i} + \vec{x}_{3,i}}{3}.$$

The contribution of each triangle to the center of gravity of the triangulated surface is proportional to its area. The area of the i-th triangle is

$$a_i = \text{area}(T_i) = \frac{|(\vec{x}_{2,i} - \vec{x}_{1,i})| \times |(\vec{x}_{3,i} - \vec{x}_{1,i})|}{2}.$$

We introduce a weight to each triangle that is proportional with its area, define as

$$w_i = \frac{a_i}{\sum_{i=1}^n a_i}.$$

Then, the center of gravity of the boundary of X is

$$\vec{c} = \sum_{i=1}^n w_i \vec{c}_i.$$

The covariance matrix of the i-th triangle is

$$\begin{aligned} C_i &= \frac{\int_0^1 \int_0^{1-s} (\vec{T}_i(s,t) - \vec{c})(\vec{T}_i(s,t) - \vec{c})^T\, dt\, ds}{\int_0^1 \int_0^{1-s} dt\, ds} \\ &= \frac{1}{12}\left(\sum_{j=1}^3 \sum_{k=1}^3 (\vec{x}_{j,i} - \vec{c})(\vec{x}_{k,i} - \vec{c})^T + \sum_{j=1}^3 (\vec{x}_{j,i} - \vec{c})(\vec{x}_{j,i} - \vec{c})^T \right). \end{aligned}$$

The element C_i^{ab} of C_i, where $a, b \in \{1, 2, 3\}$ is

$$C_i^{ab} = \frac{1}{12}\left(\sum_{j=1}^3 \sum_{k=1}^3 (x_{j,i}^a - c^a)(x_{k,i}^b - c^b) + \sum_{j=1}^3 (x_{j,i}^a - c^a)(x_{j,i}^b - c^b) \right),$$

with $\vec{c} = (c^1, c^2, c^3)$. Finally, the covariance matrix of the boundary of X is

$$C = \sum_{i=1}^{n} w_i C_i.$$

Open problem

An interesting open problem is to obtain closed form solutions for the continuous PCA over non-polyhedral objects.

Chapter 6
Experimental Results

In this chapter, we study the impact of the theoretical results from Chapter 3 and Chapter 4 on applications of several PCA variants in practice. We analyze the advantages and disadvantages of the different variants on realistic inputs, randomly generated inputs, and specially constructed (worst case) instances. For that purpose we use the closed-form solutions for continuous PCA derived in previous chapter. The results of the different PCA variants are compared with several known bounding box algorithms. The main issues of the experimental study in this chapter can be subsumed as follows:

- The traditional discrete PCA algorithm works very well on most realistic inputs. It gives a bad approximation ratio on special inputs with point clusters.

- The continuous PCA version can not be fooled by point clusters. It achieves much better approximations than guaranteed for realistic and randomly generated inputs. The only weak points come from symmetries in the input.

- To improve the performances of the algorithms we apply two approaches. First, we combine the run time advantages of PCA with the quality advantages of continuous PCA by a sampling technique. Second, we introduce a postprocessing to overcome most of the problems with specially constructed outliers.

- The thorough tests on the realistic and synthetic inputs revealed that the quality of the resulting bounding boxes was better than the guaranteed one known from the theory.

This chapter is organized as follows: In Section 6.1, we present the implementation and evaluation of several bounding box algorithms. Variants of PCA algorithms are considered in Subsection 6.1.1, and few additional bounding box algorithms are considered in Subsection 6.1.2. More detailed results of the evaluation are presented in Section 6.3. Conclusion is given in Section 6.2.

6.1 Evaluation of Bounding Box Algorithms

We have implemented and integrated in our testing environment a number of bounding box algorithms for a point set in $\mathbb{R}^3$. The algorithms were implemented using C++ and Qt, and tested on a Core Duo 2.33GHz with 2GB memory. Below we detail the algorithms used in this study. The tests were performed on real graphics models and synthetic data. The real graphics models were taken from various publicly available sources (Standford 3D scanning repository, 3d Cafe). The synthetic test data were obtained in several manners (see Figure 6.1):

- uniformly generated point set on the unit sphere;
- randomly generated point set in the unit cube;
- randomly generated clustered point set in a box with arbitrary spread.

To evaluate the influence of the clusters on the quality of the bounding boxes obtained by discrete PCA, we also generated clusters on the boundary of the real objects. The volume of a computed bounding

6.1. EVALUATION OF BOUNDING BOX ALGORITHMS

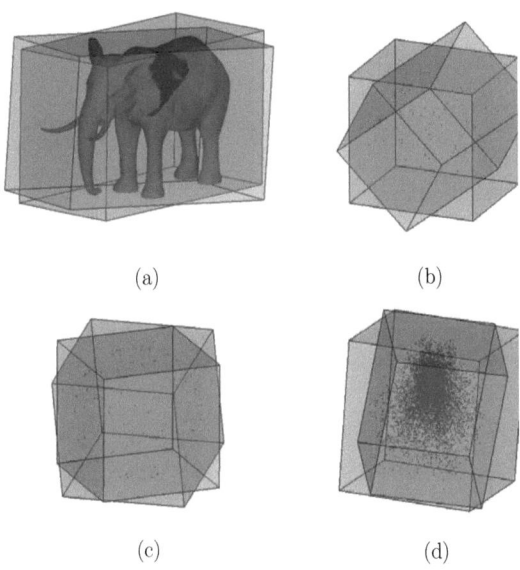

Figure 6.1: Bounding boxes of four spatial point sets: (a) real data (elephant model) (b) randomly generated point set in the unit cube (c) uniformly generated point set on the unit sphere (d) randomly generated clusters point set in a box with arbitrary dimensions.

box very often can be "locally" improved (decreased) by projecting the point set into a plane perpendicular to one of the directions of the bounding box, followed by computing a minimum-area bounding rectangle of the projected set in that plane (for example by rotating calipers algorithm [49]), and using this rectangle as the base of an improving bounding box. This heuristic converges always to a local minimum. We encountered many examples when the local minimum was not the global one. Each experiment was performed twice, with and without this improving heuristic. The parameter *#iter* in the

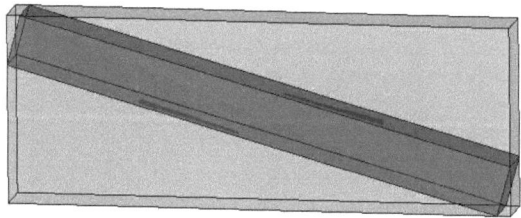

Figure 6.2: Extension of the example from Figure 3.1 in $\mathbb{R}^3$. Dense collection of additional points (the red clusters) significantly affect the orientation of the PCA bounding-box of the cuboid. The outer box is the PCA bounding box, and the inner box is the CPCA-area bounding box.

tables below shows how many times the computation of the minimum-area bounding rectangle was performed to reach a local minimum.

6.1.1 Evaluation of the PCA and CPCA Bounding Box Algorithms

We have implemented and tested the following PCA and continuous PCA bounding box algorithms:

- **PCA** - computes the PCA bounding box of a discrete point set.

- **PCA-CH** - computes the PCA bounding box of the vertices of the convex hull of a point set.

- **CPCA-area** - computes the PCA bounding box of a polyhedral surface.

- **CPCA-area-CH** - computes the PCA bounding box of the of the convex hull of a polyhedral surface.

- **CPCA-volume** - computes the PCA bounding box of a convex or a star-shaped object.

6.1. EVALUATION OF BOUNDING BOX ALGORITHMS

Figure 6.3: Igea: 134345 vertices, 268686 triangles, 5994 convex hull vertices, 11984 convex hull triangles.

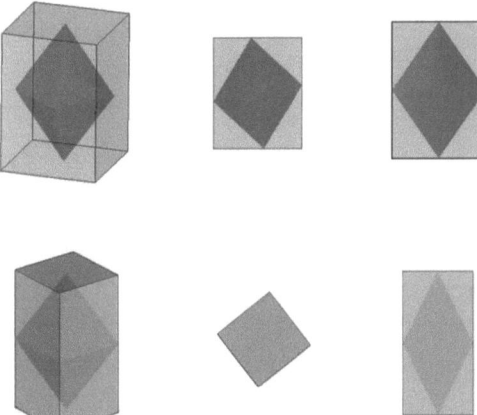

Figure 6.4: The dypiramid in the figure has two equal eigenvalues. (a) The PCA bounding box and its top and side projections. (b) The improved PCA bounding box and its top and side projections.

We have tested the above algorithms on a large number of real and synthetic objects. Typical samples of the results are given in Ta-

Table 6.1: Performance of the PCA bounding box algorithms on real data.

Igea			
algorithm	volume	#iter	time[s]
PCA	6.73373	-	0.198995
improved	6.23318	2	2.17915
PCA-CH	6.46654	-	6.16769
improved	6.22088	3	6.67348
CPCA-area	6.70684	-	0.300368
improved	6.23557	4	2.73174
CPCA-area-CH	6.72856	-	6.37164
improved	6.23379	2	6.5672
CPCA-volume	6.72727	-	5.30695
improved	6.23636	3	6.27225

ble 6.1 and Table 6.2. We give more detailed results for some of the tested data sets in the last section of this chapter. For many of the tested data sets, the volumes of the boxes obtained by CPCA algorithms were slightly smaller than the volumes of the boxes obtained by PCA, but usually the differences were negligible. However, the CPCA methods have much larger running times due to computing the convex hull. Some of the synthetic data with clusters justify the theoretical results that favors the CPCA bounding boxes over PCA bounding boxes. Figure 6.2 shows a typical example and indicates that the PCA bounding box can be arbitrarily bad.

As previously mentioned, for eigenspaces of dimension bigger than 1, the orthonormal basis of eigenvectors is chosen arbitrary. This can

6.1. EVALUATION OF BOUNDING BOX ALGORITHMS

Table 6.2: Performance of the PCA bounding box algorithms on the clustered point set with 10000 points. The values in the table are the average of the results of 100 runs of the algorithms, each time with a newly generated clustered point set.

	clustered point set		
algorithm	volume	#iter	time[s]
PCA	31.3084	-	0.036038
improved	17.4366	6	0.285556
PCA-CH	33.4428	-	1.93812
improved	17.4593	9	2.18226
CPCA-area-CH	21.0176	-	1.5961
improved	17.4559	3	1.66884
CPCA-volume	19.4125	-	1.32058
improved	17.4591	5	1.39327

result in unpredictable and large bounding boxes, see Figure 6.4 for an illustration. We solve this problem by computing bounding boxes that are aligned with one principal component. The other two directions are determined by computing the exact minimum-area bounding rectangle of the projections of the points into a plane orthogonal to the first chosen direction by rotating calipers algorithm [49].

If the connectivity of the input is known, then we can improve the run time of the PCA and PCA-area methods, without decreasing the quality of the bounding boxes, by sampling the surface and applying the PCA on the sampled points. We do the sampling uniformly, in the sense that the number of the sampled points on the particular triangle is proportional to the relative area of the triangle. Table 6.3 shows the performance of this sampling approach (denoted by PCA-sample)

Table 6.3: Performance of the sampling approach on real data. The values in the table are the average of the results of 100 runs of the algorithms, each time with a newly generated sampling point set.

Igea			
algorithm	#sampling pnts	volume	time[s]
PCA	-	6.73373	0.189644
PCA-area	-	6.70684	0.297377
PCA-sample	50	6.81354	0.122567
PCA-sample	100	6.6936	0.123895
PCA-sample	1000	6.69176	0.131753
PCA-sample	10000	6.70855	0.13825
PCA-sample	50000	6.70546	0.178306
PCA-sample	60000	6.70629	0.173158
PCA-sample	70000	6.70525	0.188299

on a real model. The results reveal that even for a small number of sampling points, the resulting bounding boxes are comparable with the PCA and CPCA-area bounding boxes. Also, if the number of the sampling points is smaller than half of the original point set the sampling approach is faster then PCA approach.

6.1.2 Evaluation of Other Bounding Box Algorithms

Next, we describe a few additional bounding box algorithms, whose performance we have analyzed.

- **AABB** - computes the axis parallel bounding box of the input point set. This algorithm reads the points only once and as such

6.1. EVALUATION OF BOUNDING BOX ALGORITHMS

Table 6.4: Performance of the additional bounding box algorithms on real data.

algorithm	volume	#iter	time[s]
\multicolumn{4}{c}{Igea}			
AABB	6.80201	-	0.008358
improved	6.36373	2	1.96749
BHP	6.45908	-	5.07754
improved	6.23635	2	7.01199
BHP-CH	6.02441	-	8.7999
improved	6.01957	1	10.3937
DiameterBB	6.6186	-	1.1151
improved	6.23595	2	2.97063

is a good reference in comparing the running times of the other algorithms.

- **BHP** - this algorithm is based on the $(1+\epsilon)$-approximation algorithm from [5], with run time complexity $O(n \log n + n/\epsilon^3)$. It is an exhaustive grid-base search, and gives far the best results among all the algorithms. In many cases, that we were able to verified, it outputs bounding boxes that are the minimum-volume or close to the minimum-volume bounding boxes. However, due to the exhaustive search it is also the slowest one.

- **BHP-CH** - same as BHP, but on the convex hull vertices.

- **DiameterBB** - computes a bounding box based on the diameter of the point set. First, a $(1-\epsilon)$ - approximation of the diameter of P that determines the longest side of the bounding box, is computed. This can be done efficiently in $O(n + \frac{1}{\epsilon^3} \log \frac{1}{\epsilon})$ time, see [18]

Table 6.5: Performance of the additional bounding box algorithms on the clustered point set with 10000 points. The results were obtained on the same point set as those from Table 6.2. A value 0, assigned to the parameter #*iter*, indicates that an algorithm has reached a local minimum and an improvement by projecting heuristic was not possible.

clustered point set			
algorithm	volume	#iter	time[s]
AABB	30.2574	-	0.000624
improved	16.4563	7	0.247101
BHP	15.5662	-	3.13794
improved	15.5662	0	3.13794
BHP-CH	15.5662	-	3.13335
improved	15.5662	0	3.13345
DiameterBB	31.5521	-	0.013173
improved	16.6952	4	0.205163

for more details, and [17] for the implementation. The diameter of the projection of P onto the plain orthogonal to longest side of the bounding box determines the second side of the bounding box. The third side is determined by the direction orthogonal to the first two sides. This idea is old, and can be traced back to [26].

Note that DiameterBB applied on convex hull points gives the same bounding box as applied on the original point set. Typical samples of the results are given in Table 6.4 and Table 6.5, for more results see the last section of this chapter.

An improvement for a convex-hull method requires less additional time than an improvement for a non-convex-hull method. This is due to the fact that the convex hull of a point set in general has fewer points than the point set itself, and once the convex hull in $\mathbb{R}^3$ is computed, it suffices to project it to the plane of projection to obtain the convex hull in $\mathbb{R}^2$. It should be observed that the number of iterations needed for the improvement of the AABB method, as well as its initial quality, depends heavily on the orientation of the point set.

6.2 Conclusion

In short, the conclusions of the experiments are as follows:

- The traditional discrete PCA algorithm can be easily fooled by inputs with point clusters. In contrast, the continuous PCA variants are not sensitive on the clustered inputs.

- The continuous PCA version on convex point sets guarantees a constant approximation factor for the volume of the resulting bounding box. However, in many applications this guarantee has to be paid with an extra $O(n \log n)$ run time for computing the convex hull of the input instance. The tests on the realistic and synthetic inputs revealed that the quality of the resulting bounding boxes was better than the theoretically guaranteed quality.

- For the most realistic inputs the qualities of the discrete PCA and the continuous PCA bounding boxes are comparable.

- The run time of the discrete PCA and continuous PCA (PCA-area) heuristics can be improved without decreasing the quality of the resulting bounding boxes by sampling the surface and applying the discrete PCA on the sampled points. This approach assumes that an input is given as a triangulated surface. If this is not the case, a surface reconstruction must be performed, which is usually slower than the computation of the convex hull.

- Both the discrete and the continuous PCA are sensitive to symmetries in the input.

- The diameter based heuristic is not sensitive to clusters and can be used as an alternative to continuous PCA approaches.

- An improvement step, performed by computing the minimum-area bounding rectangle of the projected point set, is a powerful technique that often significantly decreases the existing bounding boxes. This technique can be also used by PCA approaches when the eigenvectors are not unique.

- The experiments show that the sizes of the bounding boxes obtained by CPCA-area and CPCA-volume are comparable. This indicates that the upper bound of $\kappa_{3,2}$, which is an open problem, is probably similar to that of $\kappa_{3,3}$.

A practical and fast $(1+\epsilon)$-approximation algorithm for computing the minimum-volume bounding box of a point set in $\mathbb{R}^3$ is of general interest.

6.3 Additional results

Here, we present more detailed results of some of the tested data. In the applications, beside the volume of the bounding boxes, also other quantifiers maybe of interest. We measured also the area, the area of the biggest face and the length of the shortest edge of the bounding box, denoted in the tables as *total area, biggest face* and *shortest edge*, respectively.

6.3. ADDITIONAL RESULTS

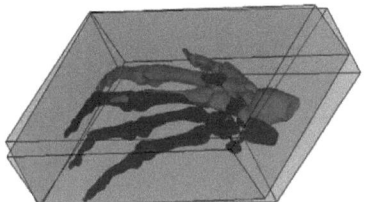

Figure 6.5: Hand: 327323 vertices, 654666 triangles, 1675 convex hull vertices, 3346 convex hull triangles and two different PCA bounding boxes - the green box is obtained with PCA-CH and the pink box is obtained with CPCA-volume algorithm.

Table 6.6: Performance of the PCA bounding box algorithms on real data.

Hand						
algorithm	volume	total area	biggest face	shortest edge		time[s]
PCA	63.5366	109.751	31.8641	1.99399	-	0.255
improved	51.2322	99.1316	30.5692	1.67594	3	5.720
PCA-CH	55.6899	104.547	32.2492	1.72686	-	6.470
improved	50.9523	99.0474	30.6574	1.66199	3	6.654
CPCA-area	64.0215	110.086	31.8564	2.00969	-	0.637
improved	51.2673	99.1355	30.5534	1.67796	3	6.021
CPCA-area-CH	54.3455	103.687	32.3132	1.68184	-	7.201
improved	50.9473	99.122	30.7141	1.65876	8	6.391
CPCA-volume	53.7146	103.209	32.3007	1.66295	-	5.313
improved	50.9485	99.1231	30.7143	1.65879	4	5.503

Figure 6.6: Lucy: 262909 vertices, 525814 triangles, 622 convex hull vertices, 1240 convex hull triangles and two different PCA bounding boxes - the green box is obtained with PCA-CH and the pink box is obtained with CPCA-volume algorithm.

Table 6.7: Performance of the PCA bounding box algorithms on real data.

algorithm	volume	total area	biggest face	shortest edge	#iter	time[s]
			Lucy			
PCA	756184000	5606960	1529520	494.392	-	0.236
improved	702004000	5381630	1497970	468.638	2	2.902
PCA-CH	736099000	5530330	1518900	484.626	-	6.470
improved	704615000	5373400	1488570	473.352	3	6.526
CPCA-area	731496000	5513090	1523000	480.3	-	0.534
improved	692082000	5327790	1484190	466.304	2	3.163
CPCA-area-CH	726545000	5471570	1504380	482.953	-	4.382
improved	696356000	5294830	1438600	484.05	2	4.474
CPCA-volume	729131000	5489320	1510740	482.632	-	5.313
improved	699059000	5305810	1442280	484.69	3	5.349

6.3. ADDITIONAL RESULTS

Figure 6.7: Buddha: 26662 vertices, 41868 triangles, 290 convex hull vertices, 576 convex hull triangles and two different PCA bounding boxes - the green box is obtained with PCA-CH and the pink box is obtained with CPCA-volume algorithm.

Table 6.8: Performance of the PCA bounding box algorithms on real data.

	Buddha					
algorithm	volume	total area	biggest face	shortest edge	#iter	time[s]
PCA	27793700	580294	136152	204.137	-	0.054
improved	21263900	507856	136112	156.224	5	0.561
PCA-CH	24465200	547630	141579	172.802	-	1.601
improved	20727700	504680	138385	149.783	4	1.270
CPCA-area	22004700	522382	142952	153.93		0.111
improved	20764200	505007	138314	150.124	4	0.549
CPCA-area-CH	22015300	523417	143706	153.197	-	1.520
improved	20719700	504608	138400	149.709	4	1.416
CPCA-volume	21934600	522443	143622	152.725	-	0.265
improved	20783500	505177	138276	150.304	5	0.310

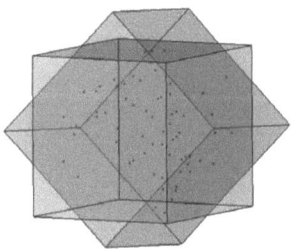

Figure 6.8: Randomly generated point set in a unit cube and two different PCA bounding boxes - the green box is obtained with PCA and the pink box is obtained with CPCA-area-CH algorithm.

Table 6.9: Performance of the additional bounding box algorithms on synthetic data (randomly generated point set in a unit cube: 70 points, 27 convex hull vertices, 53 convex hull triangles).

randomly generated point set in the unit cube						
algorithm	volume	total area	biggest face	shortest edge	#iter	time[s]
PCA	1.832	8.986	1.540	1.189	-	0.063
improved	0.943	5.771	0.980	0.962	4	0.070
PCA-CH	1.959	9.408	1.660	1.180	-	1.055
improved	0.943	5.771	0.980	0.962	5	1.267
CPCA-area-CH	1.824	8.962	1.533	1.190	-	1.204
improved	0.942	5.767	0.978	0.963	5	1.268
CPCA-volume	1.873	9.120	1.573	1.190	-	1.274
improved	0.943	5.769	0.979	0.963	5	1.293

6.3. ADDITIONAL RESULTS

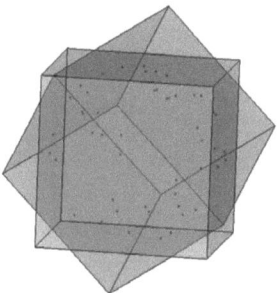

Figure 6.9: Randomly generated point set in unit sphere and two different PCA bounding boxes - the green box is obtained with PCA and the pink box is obtained with CPCA-area-CH algorithm.

Table 6.10: Performance of the additional bounding box algorithms on synthetic data (uniformly generated point set on the unit sphere: 50 points, 50 convex hull vertices, 103 convex hull triangles).

	uniformly generated point set on the unit sphere					
algorithm	volume	total area	biggest face	shortest edge	#iter	time[s]
PCA	6.92076	21.8063	3.83039	1.8068	-	0.022
improved	6.01827	19.875	3.54112	1.69954	4	0.070
CPCA-area-CH	6.69369	21.3185	3.68522	1.81636	-	2.238
improved	6.02932	19.8969	3.53324	1.70645	4	2.529
CPCA-volume	6.72143	21.3769	3.69161	1.82073	-	2.348
improved	6.02996	19.8982	3.53249	1.707	4	2.637

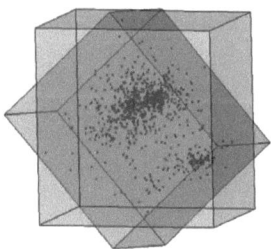

Figure 6.10: Randomly generated clustered point and two different PCA bounding boxes - the green box is obtained with PCA and the pink box is obtained with CPCA-area-CH algorithm.

Table 6.11: Performance of the additional bounding box algorithms on synthetic data (clustered point set: 10000 points, 60 convex hull points, 118 convex hull triangles).

algorithm	volume	total area	biggest face	shortest edge	#iter	time[s]
PCA	31.3084	60.3063	11.3688	2.75388	-	0.036038
improved	17.4366	41.7829	8.9374	1.95097	6	0.286
PCA-CH	33.4428	62.8968	11.6312	2.87526	-	1.93812
improved	17.4593	41.8329	8.97173	1.94603	9	2.182
CPCA-area-CH	21.0176	47.3839	10.3419	2.03227	-	1.596
improved	17.4559	41.825	8.96585	1.94693	3	1.669
CPCA-volume-CH	19.4125	45.1401	9.96599	1.94787	-	1.320
improved	17.4591	41.8326	8.97146	1.94608	5	1.393

(clustered point set)

6.3. ADDITIONAL RESULTS

Table 6.12: Performance of the additional bounding box algorithms on real data.

\multicolumn{7}{c}{Hand}						
algorithm	volume	total area	biggest face	shortest edge	#iter	time[s]
AABB	69.998	112.813	30.7663	2.27516	-	0.021
improved	51.2559	99.1346	30.5588	1.67729	7	6.982
BHP	53.1292	100.888	30.82	1.723	-	3.287
improved	51.0171	99.2622	30.7798	1.65748	1	6.773
BHP-CH	50.7504	98.6218	30.6647	1.65501	-	10.755
improved	50.7463	98.6325	30.6602	1.6549	1	10.824
DiameterBB	56.728	105.137	32.0723	1.76876	-	0.156
improved	53.0442	102.387	32.0146	1.65688	1	3.589

Table 6.13: Performance of the additional bounding box algorithms on real data.

\multicolumn{7}{c}{Lucy}						
algorithm	volume	total area	biggest face	shortest edge	#iter	time[s]
AABB	78927900	5648130	1478900	533.692	-	0.017
improved	70515200	5384180	1493950	472.006	3	4.927
BHP	74367700	5497480	1482350	501.688	-	3.252
improved	70564800	5374940	1487230	474.471	1	5.920
BHP-CH	68772300	5310360	1481590	464.179	-	8.634
improved	68769500	5310280	1481600	464.158	1	8.721
DiameterBB	150466000	8429373	2117310	710.65	-	0.123
improved	79019000	5530870	1293480	610.904	4	4.897

Table 6.14: Performance of the additional bounding box algorithms on real data.

			Buddha			
algorithm	volume	total area	biggest face	shortest edge	#iter	time[s]
AABB	23681100	537442	140661	168.356	-	0.002
improved	21260000	507914	136194	156.1	5	0.553
BHP	21311500	514305	141013	151.132	-	3.165
improved	20734900	504745	138371	149.85	3	3.580
BHP-CH	20680500	504254	138471	149.349	-	3.749
improved	20680500	504254	138471	149.349	0	4.429
DiameterBB	25876600	565151	142265	181.89	-	0.013
improved	21242800	508054	136462	155.668	8	0.556

Table 6.15: Performance of the other bounding box algorithms on synthetic data (the data set is the same as in Table 6.9).

			randomly generated point set in the unit cube			
algorithm	volume	total area	biggest face	shortest edge	#iter	time[s]
AABB	0.961049	5.84336	0.984731	0.975951	-	0.001
improved	0.943766	5.77345	0.980416	0.962618	3	0.0302
BHP	0.941845	5.76549	0.976178	0.96483	-	0.636
improved	0.941845	5.76549	0.976178	0.96483	0	0.655
BHP-CH	0.941845	5.76549	0.976178	0.96483	-	3.322
improved	0.941845	5.76549	0.976178	0.96483	0	3.437
DiameterBB	1.89185	9.19834	1.65464	1.14336	-	0.002
improved	0.94338	5.77187	0.980573	0.962069	9	0.020

6.3. ADDITIONAL RESULTS

Table 6.16: Performance of the other bounding box algorithms on synthetic data (the data set is the same as in Table 6.10).

algorithm	volume	total area	biggest face	shortest edge	#iter	time[s]
\multicolumn{7}{c}{uniformly generated point set on the unit sphere}
AABB	7.12347	22.217	3.79886	1.87516	-	0.001
improved	6.00872	19.8575	3.55337	1.69099	8	0.056
BHP	5.7772	19.3415	3.38208	1.70818	-	0.541
improved	5.7772	19.3415	3.38208	1.70818	-	0.546
DiameterBB	7.33895	22.6747	3.97946	1.84421	-	0.001
improved	5.78812	19.3657	3.38823	1.7083	4	0.034

Table 6.17: Performance of the other bounding box algorithms on synthetic data (the data set is the same as in Table 6.11).

algorithm	volume	total area	biggest face	shortest edge	#iter	time[s]
\multicolumn{7}{c}{clustered point set}
AABB	15.6724	38.1942	8.04813	1.94733	-	0.001
improved	15.6557	38.1688	8.04512	1.94599	2	0.247
BHP	15.5662	38.0147	8.00121	1.94548	-	3.138
improved	15.5662	38.0147	8.00121	1.94548	0	3.138
BHP-CH	15.5662	38.0147	8.00121	1.94548	-	3.133
improved	15.5662	38.0147	8.00121	1.94548	0	3.133
DiameterBB	31.5521	60.9057	12.0399	2.62063	-	0.013
improved	16.6952	40.0394	8.57113	1.94784	4	0.205

Chapter 7
Reflective Symmetry - an Application of PCA

7.1 Introduction and Related Work

Symmetry is one of the most important features of shapes and objects, which is proved to be a powerful concept in solving problems in many areas including detection, recognition, classification, reconstruction and matching of different geometric shapes, as well as compression of their representations. In general, symmetry in Euclidean space can be defined in terms of three transformations: translation, rotation and reflection. A subset P of $\mathbb{R}^d$ is *approximately symmetric* with respect to transformation T if for a big enough subset P' of P, the *distance* between $T(P')$ and P' is less then small constant ϵ, where the distance is measured using some appropriate metric, for example *Hausdorff, RMS (root mean square)* or *bottleneck distance measures* as most commonly used metrics. If $P' = P$ and $\epsilon = 0$, then $T(P) = P$, and we say that P is *perfectly symmetric* with respect to T. In this chapter we are interested in both approximate and perfect symmetry in terms of transformation of reflection through a hyperplane.

In what follows, we briefly survey the most relevant existing algorithms and techniques for identifying both perfect and approximate symmetry.

Traditional approaches consider perfect symmetry in discrete settings as a global feature. Some of these methods reduced the symmetry detection problem to a detection of symmetries in circular strings

[2, 19, 52, 55], for which efficient solutions are known [24]. Other efficient algorithms based on the octree representation [29], the extended Gaussian image [46] or the singular value decomposition of the points of the model [44] also have been proposed. Further, methods for describing local symmetries were developed. Blum [7] proposed an algorithm based on a medial axis transform. An algorithm presented in [48] detects perfect symmetries in range images, exploiting taxonomy of different types of symmetries and relations between them, by explicitly searching an increasing sets of points. An approach based on generalized moment functions and their spherical harmonics representation was introduced by Martinet et al. [28]. However, since the above mentioned methods consider only perfect symmetries, they may be inaccurate in detection the symmetry for shapes with added noise or missing data.

As a result to this challenge, several algorithms for measuring imperfect symmetries have been developed. For example, Zabrodsky et al. proposed an algorithm based on a *measure of symmetry*, defined as minimum mean squared distance required to transform a shape into a symmetric shape [53, 54]. A method of detecting a line of approximate symmetry of 2D images considering only the boundary of the image, using a hierarchy of certain directional codes, was presented in [37]. Marola introduced a measure of reflective symmetry with respect to a given axis where global reflective symmetry is found by roughly estimating the axis location and then fine tuning the location by minimizing the symmetry measure [27]. Kazhdan et al. introduced the *symmetry descriptors*, a collection of spherical functions that describe the measure of a model symmetry with respect to every axis passing through the center of gravity [21, 22]. Recently, Podolak et al. proposed the *planar reflective symmetry transform*, which measures

7.1. INTRODUCTION AND RELATED WORK

the symmetry of an object with respect to all planes passing through its bounding volume [40]. A method of detecting planes of reflective symmetry, by exploiting the topological configuration of the edges of a 2D sketch of a 3D objects, was developed by Zou and Lee [56]. Mitra et al. proposed a method of finding partial and approximate symmetry in 3D objects [30]. Their approach relies on matching geometry signatures (based on the concept of normal cycles) that are used to accumulate evidence for symmetries in an appropriate transformation space.

Till now, most of the research was dedicated to investigation of symmetry in 2D and 3D. Here, we consider two approaches which lead to algorithms in arbitrary dimension. The contribution of the work presented in this chapter is two-fold. First, we propose an algorithm, based on geometric hashing, for computing the reflectional symmetry of point sets with approximate symmetry in arbitrary dimension. Second, for the same purpose, we present an application of the relation between the perfect reflective symmetry and the principal components of discrete or continuous geometrical objects in arbitrary dimensions, presented in Lemma 3.1. The relation, in the case when rigid objects in 3D are considered, is known from mechanics and is established by analyzing a moment of inertia [47]. Without rigorous proof for other cases than 3D rigid objects, this result was a base as a heuristic in several symmetry detection algorithms [29, 32, 46]. Banerjee et al. also tackle this relation in 3D, in the case when the objects are represented as 3D binary arrays, but a formal proof is missing in their paper [3].

The rest of this chapter is organized as follows: In Section 7.2 we present the algorithm based on geometric hashing for computing a reflectional symmetry of a point set with approximate symmetry. The behavior of the algorithm in the 2D case is estimated by a probabilistic analysis and evaluated on real and synthetic data. In Section 7.3, we present an algorithm that exploit the relation between the perfect reflective symmetry and the principal components of geometrical objects in arbitrary dimensions. Conclusions and indications of future work are given in Section 6.2.

7.2 Detection of Reflective Symmetry: Geometric Hashing Approach

Geometric hashing is a technique originally developed in computer vision for matching geometric features against a database of such features [1, 51]. Here, we assume that the given point set $P \subseteq \mathbb{R}^d$ is approximately symmetric, and our goal is to compute the hyperplane of symmetry H_{sym} with a geometric hashing technique. More precisely, hashing is utilized to compute the normal vector of H_{sym}. Additionally, one could use the fact that the center of gravity of P lies on H_{sym} in the case when P has a perfect symmetry, or with high probability near to H_{sym} in the case when P is approximately symmetric. However, to be on the safe side, if some outliers cause that the center of gravity is far from H_{sym}, we can apply a second phase of geometric hashing to compute a point on H_{sym}.

We start from the hypothesis that each point pair (p, q) is a candidate for a pair of points that are symmetric with respect to H_{sym}. Without loss of generality, we assume that the first coordinate of p is less than or equal to the first coordinate of q. If p is symmetric to q, the vector $\overrightarrow{pq}$ is orthogonal to H_{sym}. We note that this vector is characterized uniquely by the tuple of angles $(\alpha_2, \alpha_3, \ldots, \alpha_d)$ where α_i is the angle between $\overrightarrow{pq}$ and the i-th vector of the standard base

7.2. GEOMETRIC HASHING APPROACH

of $\mathbb{R}^d$. We can omit the angle α_1 in the characterization of $\overrightarrow{pq}$ since, α_1 can be determined from the other angles. Namely, it holds that $\sin^2 \alpha_1 + \sin^2 \alpha_2 + \cdots + \sin^2 \alpha_d = 1$.

Since we assume at least a weak form of symmetry, we can expect that the number of point pairs (approximately) symmetric regarding H_{sym}, is bigger than the number of point pairs (approximately) symmetric regarding any other hyperplane H. For example, if we have a perfect symmetric point set with n points, then we have $\frac{n}{2}$ point pairs perfectly symmetric regarding H_{sym}. In contrast to that, the hyperplanes corresponding to the remaining $\binom{n}{2} - \frac{n}{2}$ point pairs are randomly distributed. See Figure 7.1 for illustration in $\mathbb{R}^2$.

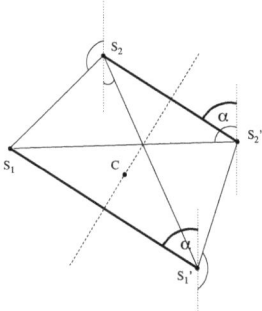

Figure 7.1: The angle α between y-axis and the line segments $\overline{s_1 s_1'}$ and $\overline{s_2 s_2'}$, formed by symmetric points, occurs two times. All other angles occurs only once.

In the standard approach of geometric hashing a number $K \in \mathbb{N}$ is fixed and the interval $[0, \pi]$ is subdivided into K subintervals of equal length π/K. Then, the hash function maps a tuple of angles $(\alpha_2, \alpha_3, \ldots, \alpha_d)$ to a tuple of integers $(a_2, a_3, \ldots, a_d)$, where each a_i denotes the index of the subinterval containing α_i, i.e.,

$$a_i = \left\lfloor \frac{\alpha_i \cdot K}{\pi} \right\rfloor.$$

Equivalently one can describe this approach with a so-called voting scheme by subdividing the cube $[0, \pi]^{d-1}$ into a grid with K^{d-1} cells. Each cell is equipped with a counter, collecting votes of all point pairs which angle tuple is contained in the cell. In the end one has to search for the cell with the maximum number of votes. However, this simple idea has some drawbacks related to the choice of K. Since K^{d-1} is a lower bound for both, time and storage complexity of the algorithm, K should not be too large. Moreover, if K is large, the noise might cause that the peak of votes is distributed over a larger cluster of cells. On the other side, if K is small, the preciseness of the result is not satisfactory.

We overcome these problems generalizing an idea from [39] that combines a rather coarse grid structure with a quite precise information about the normal vector. To this end, we use counters for the grid's vertices instead of counters for the grid's cells. Any vote $(\alpha_2, \alpha_3, \ldots, \alpha_d)$ for a grid cell $(a_2, \ldots, a_d)$ will be distributed to the incident vertices of the cell such that vertices close to $(\alpha_2, \alpha_3, \ldots, \alpha_d)$ get a larger portion of the vote than more distant vertices.

To explain this idea more precisely, we introduce some more notations. Let Q be a grid cell, and v a grid vertex incident with Q. Among the vertices incident with Q, there is exactly one, called the opposite vertex v_{opp}, that differs in all $d-1$ coordinates from v. If $\vec{\alpha} = (\alpha_2, \alpha_3, \cdots, \alpha_d)$ is a vote for Q (i.e., a point in Q) we denote by $Q(\vec{\alpha}, v)$ the (axis-parallel) subcube of Q spanned by the points $\vec{\alpha}$ and v. It is clear that the closer $\vec{\alpha}$ is to v the larger is the volume of $Q(\vec{\alpha}, v_{opp})$. Thus, the unit score of $\vec{\alpha}$ will be distributed to all vertices incident with Q such that each vertex v gets the score $vol(Q(\vec{\alpha}), v_{opp})/vol(Q)$. See Figure 7.2 for illustration of a 2-dimensional cell. We remark that K^{d-1} counters suffice for $(K+1)^{d-1}$

7.2. GEOMETRIC HASHING APPROACH

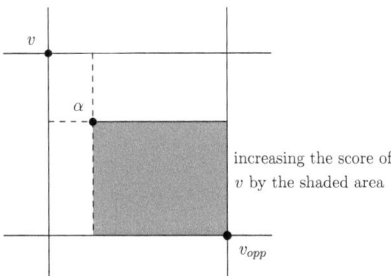

Figure 7.2: Updating the score for the angle vector $\alpha = (\alpha_1, \alpha_2)$.

grid vertices because the scoring scheme must be treated as a cyclic structure in the sense that any vertex of the form $(\beta_2, \ldots, \pi, \ldots, \beta_d)$ is identified with $(\pi - \beta_2, \ldots, 0, \ldots, \pi - \beta_d)$.

Outline of the algorithm.

Input: A set of n points $P \in \mathbb{R}^d, d \geq 2$, with approximate symmetry.
Output: An approximation of H_{sym}.

1. Let X be the set of all point pairs (p, q) from P such that the first coordinate of p is less than or equal to the first coordinate of q. Compute for each pair the angle tuple $\vec{\alpha} = (\alpha_2, \ldots, \alpha_d)$.

2. Install a voting scheme of K^{d-1} counters and set all counters to 0.

3. For each $(p, q) \in X$ with $\vec{\alpha} = (\alpha_2, \ldots, \alpha_d)$ determine the corresponding grid cell Q. For all vertices v incident with Q, add to the counter of v the vote $vol(Q(\vec{\alpha}), v_{opp})/vol(Q)$.

4. Search for the vertex $v = v_{max}$ with the largest score w. Compute the angle tuple of the approximate normal vector of H_{sym} as the

weighted center of gravity of v and its neighboring vertices with the following formula:

$$\vec{\beta} = \frac{wv + \sum_{i=2}^{d} w_i^+ v_i^+ + \sum_{i=2}^{d} w_i^- v_i^-}{w + \sum_{i=2}^{d} w_i^+ + \sum_{i=2}^{d} w_i^-},$$

where $v_i^+, v_i^-, 2 \leq i \leq d$, denote the neighboring vertices of v, and w_i^+, w_i^- their corresponding scores. Let $\vec{n}$ be a normal vector in $\mathbb{R}^d$ corresponding to the angle tuple $\vec{\beta}$.

5. Approximate a point on H_{sym} selecting all pairs $(p, q) \in X$ that vote for v_{max} (i.e., $\vec{\alpha}$ is in a cell incident with v_{max}). For each selected pair project the center $c = (p+q)/2$ onto the line spanned by the normal vector $\vec{n}$ and store the position of the projected point on that line in a 1-dimensional scoring scheme. Use the maximal score to extrapolate the location of a point on H_{sym} analogously as in 4.

Taking into account that we can keep the parameter K small, the crucial step of the algorithm is the third one, because it requires the processing of $\Theta(n^2)$ point pairs. However, it is possible to reduce this effort under the assumption that the center of gravity $c(P)$ is close to H_{sym}. This holds whenever the points without symmetric counterpart are distributed regularly in the sense that their center of gravity is close to the center of gravity of the symmetric point set. In this case it is sufficient to consider votes of pairs (p, q) of points with nearly equal distances to $c(P)$. If δ is a bound for both, the distance of $c(P)$ to H_{sym} and the distortion of the symmetric counterpart of a point with respect to H_{sym}, the first step of the algorithm can be replaced as follows:

- Compute the center of gravity $c(P)$.

- Order the points of P with respect to the distance to $c(P)$.

- For all points $q \in P$ find the first point p_i and the last point p_j in the ordered list such that $dist(p_i, c(P)) \geq dist(q, c(P)) - 2\delta$ and $dist(p_j, c(P)) \leq dist(q, c(P)) + 2\delta$ and form X from the pairs $\{q, p_k\}$, $i \leq k \leq j$.

Although this modification does not improve the run time in the worst case, it effects a remarkable speed up of the algorithm for real world data.

7.2.1 Probabilistic Analysis and Evaluation of the Algorithm in 2D Case

The 2D version of the algorithm has been implemented and tested on real and synthetic data. The generation of the synthetic data is based on a probabilistic model, which additionally can be used for a probabilistic analysis of the reliability of the algorithm.

The model incorporates the following two aspects of an approximately symmetric point set P. First, for the majority of the points $p \in P$ there is a counterpart $\tilde{p}$ that is located close to the symmetric position of p, where the symmetry, without loss of generality, is defined with respect to the x-axis. Second, there is a smaller subset of points in P without symmetric counterpart. To obtain such a point set, we apply the following procedure (see Figure 7.3 for illustration). In the upper half B^+ of the unit ball, we uniformly generate a random point set P^+ with n points. In the lower half B^- we reflect the point set P^+ over the x-axis and perturb it randomly. So, we obtain the set of points $\tilde{P}^- = \{(x \pm \delta_x, -y \pm \delta_y) \mid (x,y) \in P^+\}$, where (δ_x, δ_y) is random point from the ball $B((0,0), \epsilon)$. Additionally, we generate a random point set M in B, with m points, which do not have symmetric counterpart. Point set M represents an additional noise in the

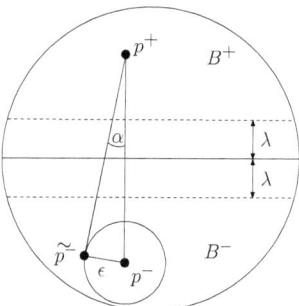

Figure 7.3: Point set generation.

form of missing/extra points in the input data set.

Most pairs of symmetric points span a line that is nearly parallel to the y-axis. A vote of such pair will be called a *good* vote. Nevertheless, for points $p^+ \in P^+$ that are close to the x-axis the perturbation of p^- might cause a bigger angle α between the y-axis and the line spanned by $\widetilde{p^-}$ and p^+. A vote from such point pairs, as well as votes from nonsymmetric point pairs, will be called *bad*. Thus, we introduce a parameter $\lambda > 0$ defining a stripe of width 2λ along the x-axis such that all symmetric point pairs out of this stripe have good votes.

Our goal is to derive an upper bound for ϵ that makes almost sure, that the given symmetry line corresponds to a maximal peak in the scoring scheme. We first estimate the width of the interval collecting the votes of the majority of the correct point pairs regarding to the symmetry line. On the other side, we will show that the probability, that another interval of the the same width would collect the same order of votes, is very small for bounded ϵ.

Since the scoring scheme is a cyclic structure, it also makes sense to speak about negative angles: especially, angles $\alpha \in (\frac{\pi}{2}, \pi)$ will be identified with the negative angles $\alpha - \pi \in (-\frac{\pi}{2}, 0)$. According to

7.2. GEOMETRIC HASHING APPROACH

Figure 7.3, for a symmetric point pair outside the λ stripe we have the following bound on the angle α which defines the vote of the pair: $\sin\alpha \leq \frac{\epsilon}{2\lambda}$, or $|\alpha| \leq \arcsin\frac{\epsilon}{2\lambda}$. Since $\arcsin\frac{\epsilon}{h} \leq \frac{\pi}{2}\frac{\epsilon}{h} \leq \frac{\pi}{2}\frac{\epsilon}{2\lambda}$, we have

$$|\alpha| \leq \arcsin\frac{\epsilon}{2\lambda} \leq \frac{\pi\epsilon}{4\lambda}. \quad (7.1)$$

We set $\gamma(\epsilon, \lambda) := \frac{\pi\epsilon}{4\lambda}$ and introduce for any angle β the random variable V_β counting all votes of the random point set P that fall into the interval $[\beta - \gamma(\epsilon, \lambda), \beta + \gamma(\epsilon, \lambda)]$.

Let $A_1 = \pi/2$ denote the area of the upper half of the unit ball and $A_2 = 2\lambda$ denote the area of the rectangle over the horizontal diameter of the unit ball with height λ. Thus, the probability that a point $p \in P^+$ generates a good pair is at least $q = \frac{A_1 - A_2}{A_1} = (1 - \frac{4\lambda}{\pi})$. Since V_0 is at least the sum S of n independent variables

$$X_i = \begin{cases} 1 & \text{with probability } q; \\ 0 & \text{with probability } 1 - q, \end{cases}$$

we have

$$E(V_0) \geq E(S) = nq, \quad (7.2)$$

and

$$Pr[V_0 < t] \leq Pr[S < t], \quad \forall t > 0. \quad (7.3)$$

Combining (7.3) with the Chernoff inequality $Pr[S < E[S] - t] \leq e^{-2t^2/n}$, for $t = E[S]/2 = nq/2$, we obtain the following estimation:

$$Pr[V_0 < nq/2] \leq e^{-q^2 n/2}. \quad (7.4)$$

Let $N \leq \binom{2n+m}{2}$ be the number of points pairs with bad votes, and consider an angle β, where $|\beta| > 2\gamma(\epsilon, \lambda)$, i.e., X_β does not count any good vote. The expectation of X_β is

$$E(X_\beta) = N\frac{2\gamma(\epsilon, \lambda)}{\pi} = N\frac{\epsilon}{2\lambda}. \quad (7.5)$$

Applying the Markov inequality $Pr[X_\beta > t] \leq \frac{E(X_\beta)}{t}$, for $t = nq/2$, we obtain

$$Pr[X_\beta > nq/2] \leq \frac{N\epsilon}{\lambda q n}. \tag{7.6}$$

We would like to note that in the case of X_β, we cannot apply any of the Chernoff's inequalities, which in general give better bounds than the Markov inequality, because X_β is not a sum of independent random variables.

Now, we come to the ultimate goal of this analysis - to estimate $Pr[V_\beta > V_0]$ and to study when it is small, i.e., when the algorithm gives a correct answer with high probability. Combining

$$Pr[V_\beta > V_0] \leq Pr[V_\beta > t] + Pr[V_0 < t], \quad \forall t > 0, \tag{7.7}$$

with (7.4) and (7.6), we obtain

$$Pr[V_\beta > V_0] \leq e^{-q^2 n/2} + \frac{N\epsilon}{\lambda q n}. \tag{7.8}$$

The first term of the right side of (7.8) is significantly smaller then the second term. This can be explained by the fact that the first term was obtained by the Chernoff inequality, and the second term by the weaker Markov inequality. However, for $\epsilon = o(\frac{1}{n})$ the second term will be also small, and then the algorithm will work well with high probability.

We implemented the algorithm and evaluated its performance on a Intel-core 2 duo computer with 2GB central memory. As described above, we randomly generated 100 point sets with same parameters ϵ and k, where k is the ratio between the number of additional points and the number of good point pairs ($k = m/n$). Table 7.1 shows the empirical probability of finding the correct angle of the symmetry line. We present here only those combination of ϵ and k for which the empirical probability was at least 0.9. The results indicate that the

7.2. GEOMETRIC HASHING APPROACH

Table 7.1: Empirical probability of finding correct line of reflective symmetry for different values of the "noise" parameters ϵ and k.

$k \setminus \epsilon$	0.01	0.005	0.004	0.003	0.002	0.001	0.0
0.9	0.90	0.92	0.93	0.94	0.94	0.95	0.95
0.8	0.91	0.93	0.94	0.95	0.95	0.96	0.96
0.7	0.91	0.93	0.94	0.94	0.95	0.96	0.97
0.6	0.94	0.93	0.96	0.96	0.97	0.99	0.99
0.5	0.96	0.99	0.96	0.99	1.0	1.0	1.0
0.4	1.0	1.0	1.0	1.0	1.0	1.0	1.0
0.3	1.0	1.0	1.0	1.0	1.0	1.0	1.0
0.2	1.0	1.0	1.0	1.0	1.0	1.0	1.0
0.1	1.0	1.0	1.0	1.0	1.0	1.0	1.0
0.0	1.0	1.0	1.0	1.0	1.0	1.0	1.0

algorithms is less sensitive to noise, due to missing/extra data, then to noise that comes from imperfect symmetry of the points. This conclusion is consistent with the theoretical analysis we have obtained. Namely, ϵ and N occur at the same place in the last term of the relation (7.8). The number of additional points m occurs in the relation (7.8) through N. The other variable which determines N is n, and its contribution to the value of N is bigger than that of m. Therefore, m has smaller influence to the expression than ϵ. We tested the algorithm also on real data sets. The tests were performed on pore patterns of copepods - a group of small crustaceans found in the sea and nearly every freshwater habitat (see Figure 7.4). The pores in a pattern were detected as points by the method based on a combination of hierarchical watershed transformation and feature extraction methods presented in [39]. The algorithm successfully detected the symmetry line because the extracted point sets have relatively good reflective symmetry, and majority of the points (around 90%) have a symmetric counterpart.

An implementation of the geometric hashing algorithm in higher dimensions and estimations of its behavior is of future interest. Of course, the 3D case is of the biggest practical importance.

7.3 Detection of Reflective Symmetry: PCA Approach

Another approach for an efficient detection of the hyperplane of perfect reflective symmetry in arbitrary dimension is that based on principal component analysis [16]. To the best of our knowledge, this approach was used as heuristic without rigorous proof (also confirmed in communication with other researchers in this area [33]). A relation between the principal components and symmetry of an object, in the case of rigid objects in 3D, was establish in mechanics by analyzing a moment of inertia [47]. This result, in the context of detecting the symmetry, was first exploit by [29]. Here, we use Lemma 3.1, that extends that result to any set of points (continuous or discrete) in arbitrary dimension.

As an immediate consequence of Lemma 3.1 we have:

Corollary 7.1. *Let P be a perfectly symmetric point set in arbitrary dimension. Then, any hyperplane of reflective symmetry is spanned by n-1 principal axes of P.*

The corollary implies a straightforward algorithm for finding the hyperplane of reflective symmetry of a point set in arbitrary dimen-

7.3. PCA APPROACH

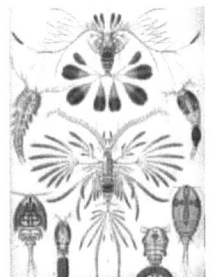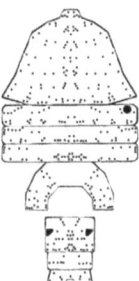

Figure 7.4: Left side: illustrations of different types of copepods. Right side: a pore pattern of a copepod.

sion.

Outline of the algorithm.

Input: A set of n points $P \in \mathbb{R}^d, d \geq 2$, with approximate symmetry.
Output: An approximation of H_{sym}.

1. Compute the covariance matrix C of P.

2. Compute the eigenvectors of C and the candidate hyperplanes of reflective symmetry.

3. Reflect the points through every candidate hyperplane.

4. Find if each reflected point is close enough to a point in P. The correspondence between reflected points and points in P is bijection.

The first and third step of the algorithm have linear time complexity in the number of points. Computation of the eigenvectors, when d is not very large, can be done in $O(d^3)$ time, for example with Jacobi or QR method [41]. Computing the candidate hyperplanes can be done in $O(d)$. Therefore, for fixed d, the time complexity of the second step is constant. For very large d, the problem of computing eigenvalues is non-trivial. In practice, the above mentioned methods for computing eigenvalues converge rapidly. In theory, it is unclear how to bound the running time combinatorially and how to compute the eigenvalues in decreasing order. In [9] a modification of the *Power method* [36] is presented, which can give a guaranteed approximation of the eigenvalues with high probability. However, for reasonable big d the most expensive step is the forth one. Here we can apply an algorithm for nearest neighbor search, for example the algorithm based on Voronoi diagram, which together with preprocessing has run time complexity $O(n \log n)$, $d = 2$, or $O(n^{\lceil \frac{d}{2} \rceil})$, $d \geq 3$. If we consider point sets with perfect symmetry, then in the 4-th step, it suffices to check if the reflection of a point of P is identical with another point of P. For example, this can be done by sorting the points with respect to each of d coordinates, and applying binary search. For the same purpose, more sophisticated data structure can be used, but however, the complexity to build them will be predominate by the factor $O(d\, n \log n)$ that comes from the sorting of the point set.

In what follows, we discuss two problems that may arise in theory, but are relatively uncommon in practice. Both were already mentioned in Chapter 2. Here, we will consider them in the context of determining reflective symmetry. The first one considers the case when the eigenvalues are not distinct, and the other the case when one or more variances are zero.

Equality of eigenvalues, and hence equality of variances of PCs, will occur for certain distribution of points. The effect of this occurrence is that for a group of q equal eigenvalues, the corresponding q eigenvectors span a certain unique q-dimensional space, but, within this space, they are, apart from being orthogonal to one another, arbitrary. In

7.3. PCA APPROACH

the context of our problem, it means that the d-dimensional point set will have exactly d candidates as hyperplanes of symmetry only when the eigenvalues of the covariance matrix are distinct. For example, if we have 3-dimensional point set, then if exactly two eigenvalues of the covariance matrix are equal, than the point set might has rotational and reflective symmetry. Such an example is the dypiramid from Figure 6.4. If the all three eigenvalues are equal, the point set might have any type of symmetry, including spherical symmetry. This is a case, for example, of a point set consisting of corner-points of a cube. In the case when the eigenvalues are not distinct, we can slightly perturb the point set, and obtain unique approximate hyperplanes of reflective symmetry.

The case when q variances equal zero, implies that the rank of the covariance matrix of the point set diminishes for q. Therefore we can reduce the d-dimensional problem to a $(d-q)$-dimensional problem.

We tested the PCA algorithm on the same data set and same computer as the geometric hashing algorithm. As it was expected, the preciseness of the PCA algorithm was inferior for data with imperfect symmetry. It fails to detect correctly the symmetry line even for small values of the parameters k and ϵ, for which the geometric hashing algorithm always find the correct solutions. However, for data with perfect, or almost prefect symmetry, the PCA algorithm is a better choice, since it is much faster. For example, for a set of 10000 points, the PCA algorithms takes 0.1 second, while the geometric hashing algorithm needs about 25 seconds. This was also expected, since the first algorithm has linear and the second quadratic run time.

Thus, we conclude, that beside its simplicity and efficiency, detecting symmetry by PCA has two drawbacks. PCA fails to identify potential hyperplanes of symmetry, when the eigenvalues of the covariance matrix of the object are not distinct. The second drawback is that the PCA approach cannot guaranty the correct identification when the symmetry of the shape is weak.

Bibliography

[1] H. Alt, and Guibas, L.. Discrete geometric shapes: Matching, interpolation, and approximation. In Sack, J.-R. and Urrutia, J., editors, *Handbook of Computational Geometry*, pages 121 – 153. Elsevier Science Publishers B.V. North-Holland, Amsterdam, 1999.

[2] M. J. Atallah. On symmetry detection. *IEEE Transactions on Computers*, 34(7):663–666, 1985.

[3] D. K. Banerjee, S. K. Parui, and D. D. Majumder. Plane of symmetry of 3d objects. In *Proceedings of the 29th Annual Convention of the Computer Society of India. Information Technology for Growth and Prosperity.*, pages 39–44, 1994.

[4] G. Barequet, B. Chazelle, L. J. Guibas, J. S. B. Mitchell, and A. Tal. Boxtree: A hierarchical representation for surfaces in 3D. *Computer Graphics Forum*, (15):387–396, 1996.

[5] G. Barequet, and S. Har-Peled. Efficiently approximating the minimum-volume bounding box of a point set in three dimensions. In *Journal of Algorithms*, 38(1): 91–109, 2001.

[6] N. Beckmann, H.-P. Kriegel, R. Schneider, and B. Seeger. The R^*-tree: An efficient and robust access method for points and rectangles. In *Proceedings of ACM SIGMOD International Conference on Management of Data*, pages 322–331, 1990.

[7] H. Blum. A transformation for extracting new descriptors of shape. In *Models for the Perception of Speech and Visual Form*, pages 362–380, 1967. MIT Press, W. Whaten-Dunn, Ed.

[8] C. Chatfield, A. J. Collins, and J. M. Bibby. *Introduction to Multivariate Analysis*. Chapman and Hall, London, 1980.

[9] S.-W. Cheng, Y. Wang, and Z. Wu. Provable dimension detection using principal component analysis. In *Proceeding of the 21st ACM Symposium on Computational Geometry*, pages 208–217, 2005.

[10] D. Dimitrov, M. Holst, C. Knauer, and K. Kriegel. Experimental study of bounding box algorithms. In *Proceedings of International Conference on Computer Graphics Theory and Applications - GRAPP 2008*, pages 15–22, 2008.

[11] D. Dimitrov, C. Knauer, K. Kriegel, and G. Rote. Upper and lower bounds on the quality of the PCA bounding boxes. In *Proceeding of International Conference in Central Europe on Computer Graphics, Visualization and Computer Vision - WSCG 2007*, pages 185–192, 2007.

[12] D. Dimitrov, C. Knauer, K. Kriegel, and G. Rote. New upper bounds on the quality of the PCA bounding boxes in $\mathbb{R}^2$ and $\mathbb{R}^3$. In *Proceedings of the 23rd ACM Symposium on Computational Geometry* , pages 275–283, 2007.

[13] D. Dimitrov, and K. Kriegel. Detection of perfect and approximate reflective symmetry in arbitrary dimension. In *Proceedings of International Conference on Computer Vision Theory and Applications - VISAPP 2007* , pages 128–136 , 2007.

[14] S. Gottschalk, M. C. Lin, and D. Manocha. OBBTree: A hierarchical structure for rapid interference detection. *Computer Graphics*, 30:171–180, 1996.

[15] B. Grünbaum. Partitions of mass-distributions and convex bodies by hyperplanes. *Pacific Journal of Mathematics*, 10:1257–1261, 1960.

[16] I. Jolliffe. *Principal Component Analysis*. Springer-Verlag, New York, 2nd ed., 2002.

[17] S. Har-Peled. Source code of program for computing and approximating the diameter of a point-set in 3d. *http://www.uiuc.edu/~sariel/papers/00/diameter/diam_prog.html*, 2000.

[18] S. Har-Peled. A practical approach for computing the diameter of a point set. In *Proceedings of the 17th ACM Symposium on Computational Geometry*, pages 177–186, 2001.

[19] P. T. Highnam. Optimal algorithms for finding the symmetries of a planar point set. *Information Processing Letters*, 22(5):219–222, 1986.

[20] H. Hotelling. Analysis of a complex of statistical variables into principal components. *Journal of Educational Psychology*, 24, pages 417–441, 498-520, 1933.

[21] M. Kazhdan, B. Chazelle, D. Dobkin, T. Funkhouser, and S. Rusinkiewicz. A reflective symmetry descriptor for 3d models. *Algorithmica*, 38:201–225, 2003.

[22] M. Kazhdan, T. Funkhouser, and S. Rusinkiewicz. Symmetry descriptors and 3D shape matching. In *Proceedings of Symposium on Geometry Processing*, pages 115–123, 2004.

[23] H. Kharaghani, and B. Tayfeh-Rezaie. A Hadamard matrix of order 428. *Journal of Combinatorial Designs 13*, pages 435–440, 2005.

[24] D. Knuth, J. H. Morris, and V. Pratt. Fast pattern matching in strings. *SIAM Journal on Computing*, 6(2):323–350, 1977.

[25] M. Lahanas, T. Kemmerer, N. Milickovic, D. B. K. Karouzakis, and N. Zamboglou. Optimized bounding boxes for three-dimensional treatment planning in brachytherapy. *Medical Physics*, 27:2333–2342, 2000.

[26] A. M. Macbeath. A compactness theorem for affine equivalence classes of convex regions. *Canadian Journal of Mathematics*, 3:54–61, 1951.

[27] G. Marola. On the detection of the axes of symmetry of symmetric and almost symmetric planar images. *IEEE Transactions on Pattern Analysis and Machine Intelligence*, 11(1):104–8, 1989.

[28] A. Martinet, C. Soler, N. Holzschuch, and F. Sillion. Accurate detection of symmetries in 3d shapes. *ACM Transactions on Graphics*, 25(2):439–464, 2006.

[29] P. Minovic, S. Ishikawa, and K. Kato. Symmetry identification of a 3-d object represented by octree. *IEEE Transaction on Pattern Analysis and Machine Intelligence*, 15(5):507–514, 1993.

[30] N. J. Mitra, L. Guibas, and M. Pauly. Partial and approximate symmetry detection for 3d geometry. *ACM Transactions on Graphics*, 25:560–568, 2006.

[31] B. Mityagin. Two inequalities for volumes of convex bodies. *Mathematical Notes*, 5:61–65, 1968.

[32] D. O'Mara, and R. Owens. Measuring bilateral symmetry in digital images. In *Proceedings o IEEE TENCON Digital Signal Processing Application*, pages 151–156, 1996.

[33] D. O'Mara, and R. Owens. Private communication, 2005.

[34] K. V. Mardia, J. T. Kent, and J. M. Bibby. *Multivariate Analysis*. Academic Press Inc., New York, 1980.

[35] J. O'Rourke. Finding minimal enclosing boxes. *International Journal of Computer and Information Science*, 14:183–199, 1985.

[36] B. N. Parlett. *The Symmetric Eigenvalue Problem*. Society of Industrial and Applied Mathematics (SIAM), Philadelphia,PA, 1998.

[37] S. Parui, and D. Majumder. Symmetry analysis by computer. *Pattern Recognition*, 16:63–67, 1983.

[38] K. Pearson. On lines and planes of closest fit to systems of points in space. *Philosophical Magazine*, 6(2):559–572, 1901.

[39] K.-P. Pleißner, F. Hoffmann, K. Kriegel, C. Wenk, S.Wegner, A. Sahlströhm, H. Oswald, H. Alt, and E. Fleck. New algorithmic approaches to protein spot detection and pattern matching in two-dimensional electrophoresis gel databases. *Electrophoresis*, 20:755–765, 1999.

[40] J. Podolak, P. Shilane, A. Golovinskiy, S. Rusinkiewicz, and T. Funkhouser. A planar-reflective symmetry transform for 3D shapes. *ACM Transactions on Graphics*, 25(3):549–559, 2006.

[41] W. H. Press, S. A. Teukolsky, W. T. Veterling, and B. P. Flannery. *Numerical Recipes in C: The art of scientific computing*. Cambridge University Press, New York, USA, second edition, 1995.

[42] N. Roussopoulos, and D. Leifker. Direct spatial search on pictorial databases using packed R-trees. In *ACM SIGMOD*, pages 17–31, 1985.

[43] T. Sellis, N. Roussopoulos, and C. Faloutsos. The R^+-tree: A dynamic index for multidimensional objects. In *Proceedings of the 13th Very Large Data Bases (VLDB) Conference*, pages 507–518, 1987.

[44] M. I Shah, and D. C. Sorensen. A symmetry preserving singular value decomposition. *SIAM Journal of Matrix Analysis and Applications*, 28:749–769, 2005.

[45] T. Siotani, Y. Fujikoshi, and T. Hayakawa. *Modern multivariate statistical analysis, a graduate course and handbook*. American Sciences Press, Syracuse, New York, 1986.

[46] C. Sun, and J. Sherrah. 3d symmetry detection using the extended gaussian image. *IEEE Transactions on Pattern Analysis and Machine Intelligence*, 19(2):164–168, 1997.

[47] K. R. Symon. *Mechanics*. Addison-Wesley, Reading, Massachusetts, 3rd edition, 1971.

[48] S. Thrun, and B.Wegbreit. Shape from symmetry. In *Proceedings of International Conference on Computer Vision (ICCV)*, pages 1824–1831. IEEE Computer Scoiety, 2005.

[49] G. Toussaint. Solving geometric problems with the rotating calipers. In *Proceedings of the 2nd IEEE Mediterranean Electrotechnical Conference (MELECON)*, May 1983.

[50] D. V. Vranić, D. Saupe, and J. Richter. Tools for 3D-object retrieval: Karhunen-Loeve transform and spherical harmonics. In

Proceeding of the IEEE 2001 Workshop Multimedia Signal Processing, pages 293–298, 2001.

[51] H. Wolfson, and I. Rigoutsos. Geometric hashing: An overview. IEEE Computational Science and Engineering 4:10–21, 1997.

[52] J. D. Wolter, T. C. Woo, and R. A. Volz. Optimal algorithms for symmetry detection in two and three dimensions. *The Visual Computer 1*, 1:37–48, 1985.

[53] H. Zabrodsky, S. Peleg, and D. Avnir. Completion of occluded shapes using symmetry. In *Proceedings of Conference on Computer Vision and Pattern Recognition*, pages 678–679, 1993.

[54] H. Zabrodsky, S. Peleg, and D. Avnir. Symmetry as a continuous feature. *IEEE Transactions on Pattern Analysis and Machine Intelligence*, 17(12):1154–1166, 1995.

[55] J. Zhang, and K. Huebner. Using symmetry as a feature in panoramic images for mobile robot applications. In *Proceedingsof Robotik*, VDI-Berichte, volume 1679, pages 263–268, 2002.

[56] H. L. Zou, and Y. T. Lee. Skewed mirror symmetry detection from a 2d sketch of a 3d model. In *GRAPHITE '05: Proceedings of the 3rd international conference on Computer graphics and interactive techniques in Australasia and South East Asia*, pages 69–76. ACM Press, 2005.

i want morebooks!

Buy your books fast and straightforward online - at one of world's fastest growing online book stores! Environmentally sound due to Print-on-Demand technologies.

Buy your books online at
www.get-morebooks.com

Kaufen Sie Ihre Bücher schnell und unkompliziert online – auf einer der am schnellsten wachsenden Buchhandelsplattformen weltweit! Dank Print-On-Demand umwelt- und ressourcenschonend produziert.

Bücher schneller online kaufen
www.morebooks.de

 VDM Verlagsservicegesellschaft mbH
Heinrich-Böcking-Str. 6-8   Telefon: +49 681 3720 174   info@vdm-vsg.de
D - 66121 Saarbrücken   Telefax: +49 681 3720 1749   www.vdm-vsg.de

Printed by Books on Demand GmbH, Norderstedt / Germany